Section 3-7 You can solve projectile motion problems using techniques learned for straight-line motion

Goal: Solve problems involving projectile motion.

Concept Check
1. If a pellet gun is fired horizontally and at the same instant a pellet is dropped from the same height, which will hit the ground first?

Problem-Solving Review
A cannon is fired from the ground at an angle of 27.0° from the horizontal with an initial velocity of 56.0 m/s.

Set Up
1. Find v_{0x} and v_{0y}.

2. Fill out the Know/Don't Know table below using the information given in the problem and your calculations in question 1.

Equation	Quantity										
	t	x	x_0	v_{0x}	v_x	a_x	y	y_0	v_{0y}	v_y	a_y
$v_x = v_{0x} + a_x t$											
$x = x_0 + v_{0x} t + \frac{1}{2} a_x t^2$											
$v_y = v_{0y} + a_y t$											
$y = y_0 + v_{0y} t + \frac{1}{2} a_y t^2$											

3. Write the equation will you use to find the time it takes for the cannonball to reach its highest height.

4. Write the equation you will use to determine the distance the cannonball travels.

Solve

1. What is the highest height the cannonball reaches?

2. How far does the cannonball go?

3. What is the final velocity of the cannonball?

4. How long is the cannonball in the air?

Reflect

1. Does the cannonball spend more time going up or down?

Section 3-8 An object moving in a circle is accelerating even if its speed is constant

Goal: Describe why an object moving in a circle is always accelerating.

Concept Check
1. Explain why an object moving in a circle with a constant tangential velocity is accelerating.

Problem-Solving Review
A car moving at a constant speed of 120 km/hr travels straight for 1.00 km and then rounds a circular bend. As the car rounds the bend it experiences a centripetal acceleration of 40.0 m/s². What is the radius of curvature of the bend?

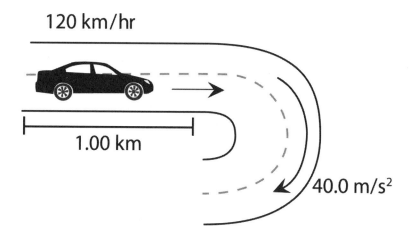

Set Up
1. On the picture above draw the velocity and acceleration vectors of the car when it is on the straight section and when it is in the middle of the curve.

Chapter 3

2. Is the car accelerating while it is traveling on the straight parts of the track? Is it accelerating around the bend?

Solve

1. Write an expression for the centripetal acceleration of the car as it rounds the bend in terms of its velocity and the radius of curvature of the bend (just use letters...no numbers yet!).

2. Solve the expression you found in question 1 for the radius of curvature of the bend.

Reflect

1. If you were a passenger in the car would you feel like you were being pulled inward or pushed outward as the car rounds the bend?

Section 3-9 Any problem that involves uniform circular motion uses the idea of centripetal acceleration

Goal: Solve problems that involve motion in a circle.

Concept Check
1. Does an airplane experience circular motion? If so, what is it circling about?

Problem-Solving Review
A typical commercial aircraft travels at 500 mi/hr at around 35,000 ft above sea level.

Set Up
1. Sketch the plane and draw the acceleration and velocity vector for the plane's motion.

2. What is the distance from the plane to the center of the Earth in meters? (The radius of the Earth is 6378 km.)

Solve
1. Find the centripetal acceleration of the plane.

2. For an object to be in orbit, the centripetal acceleration of the object would have to equal the gravitational acceleration. How fast would the plane have to be going at its current altitude for this to be true?

Reflect

1. If you kept the planes speed constant, would you increase or decrease its altitude to put it in orbit?

2. Explain why it is not possible for the plane to be in orbit at 500 mi/hr.

3. Why would you feel weightless on the space station, but not on an airplane?

Section 3-10 The vestibular system of the ear allows us to sense acceleration

Goal: Explain how the structure of the ear helps us sense acceleration.

Concept Check

1. Briefly explain how the vestibular system helps us sense acceleration.

2. Why do we get seasick?

Chapter 4 Forces and Motion I: Newton's Laws

Section 4-1 How objects move is determined by the force that act on them

Goal: Describe the importance of forces in determining how an object moves.

Concept Check

1. A tugboat is pulling a large oil tanker at constant velocity with a force of magnitude F_A. How large is the force that the tanker exerts on the tugboat? What is the direction of this force?

2. There is a great amount of drag on both the tugboat and tanker due to the water surrounding them. Which force determines the motion of the tanker: F_A, the drag force; F_B, the force of the tugboat; or F_C, the force of the tugboat minus the drag force? Explain how you determined the answer.

Section 4-2 If a net external force acts on an object, the object accelerates

Goal: Use Newton's second law to relate the net force on an object to the object's acceleration.

Concept Check

1. An object experiences two forces: F_A acting in the $+x$ direction and F_B acting in the $-x$ direction. If the magnitude of F_B is greater than F_A, which direction is the net force? Is there an acceleration? If the mass of the object were increased, what would happen to the acceleration?

2. On the way to go sledding, you pull your sled behind you at a constant velocity. What is the net force exerted on the sled? What is the acceleration?

Problem-Solving Review

At summer camp a group of kids tie two ropes to a 25.0 kg kettlebell so that they can play tug of war with four teams. Team A pulls their end of the rope with 510 N in the $+x$ direction, while Team B pulls in the exact opposite direction with 379 N. Team C exerts a force of 393 N in a direction 33° above Team A's rope. Team D pulls in the exact opposite direction of Team C with a force of 543 N.

Set Up

1. Sketch the forces acting on the kettlebell.

2. Find the *x* components of the forces acting on the kettlebell.

3. Find the *y* components of the forces acting on the kettlebell.

Solve

1. Find the *x* and *y* components of the net force on the kettlebell. You can find the *x* component of the net force by summing the *x* components of each team's forces, and likewise for the *y* component.

2. Use Newton's second law to find the *x* and *y* components of the kettlebell's acceleration.

3. What is the magnitude and direction of the kettlebell's acceleration?

4. How long does it take for the kettlebell to get 4.00 m from where it started?

Reflect

1. What would the magnitude and direction of the kettlebell's acceleration be if the mass were twice as large?

Section 4-3 Mass, weight, and inertia are distinct but related concepts

Goal: Recognize the distinctions among mass, weight, and inertia.

Concept Check
1. A bowling ball and a tennis ball roll off a table. Ignoring friction and air resistance, which ball will hit the ground first? On which ball is the force of gravity the greatest? Which ball has the greater gravitational acceleration? How do you know?

2. An astronaut in deep space (where gravity is negligible) throws a ball. What are the forces acting on the ball once it leaves the astronauts hand? Assuming it doesn't bump into anything, how far will it go? Explain your answer.

Problem-Solving Review
You are helping a friend move into a new apartment. To get the dresser down the street to the moving truck, you push it from behind with 250 N while your friend pulls it with 200 N. After you get it started the dresser, which weighs 500 N, moves at a constant velocity.

Set Up
1. What is the mass of the dresser?

2. Draw a diagram of the forces acting on the dresser.

3. What are the forces acting on the dresser?

Solve
1. What is the magnitude of the net force on the dresser?

2. If the *x* direction is parallel to the ground, which forces are acting in the *x* direction? What about the *y* direction?

3. Use Newton's first law to write an equation for the forces in the *x* direction.

4. Solve the expression you found in question 3 above for the force of friction on the dresser.

Reflect
1. As you get closer to the moving truck you push harder on the dresser, so that it is kept at a constant acceleration of 0.750 m/s². What is the magnitude of the net force acting on the dresser?

Forces and Motion I

Section 4-4 Making a free-body diagram is essential in solving any problem involving forces

Goal: Draw and use free-body diagrams in problems that involve forces.

Concept Check
1. Why is it important to draw free-body diagrams?

Problem-Solving Review
Consider the following three cases:

Case A
A box resting on a ramp at an angle θ with no friction.

Case B
One block on a table connected to another block hanging off the edge by a string and a massless pulley.

Case C
Two blocks on a table connected by a string. The first block is being pulled on in the $-x$ direction.

Set Up
1. Sketch Case A, Case B, and Case C. Be sure to specify the directions of the x and y axes.

Solve

1. Draw the free-body diagram for Case A. Be sure to specify the x and y axes.

2. Draw the free-body diagram for each of the blocks in Case B. Be sure to specify the x and y axes.

3. Draw the free-body diagram for each of the two blocks in Case C. Be sure to specify the x and y axes.

Reflect

1. Why do you need to draw a free-body diagram for each of the blocks in Case B and Case C?

2. For Case A, why is it advantageous to choose the *x* and *y* axes along the incline?

3. In all cases, the systems will accelerate. Why is there no $m\vec{a}$ in their free-body diagrams?

Chapter 4

Section 4-5 Newton's third law relates the forces that two objects exert on each other

Goal: Describe how Newton's third law relates forces that act on different objects.

Concept Check

1. You hold a tennis ball in your hand such that it is not moving. Earth exerts a force of 5.00 N on the ball pulling it downward. How much force is the ball applying to your hand? How much force is your hand applying to the ball? How much force is the ball applying to Earth?

2. Consider a person pushing on a box so it is accelerating with acceleration $\vec{a}$. Draw a free-body diagram for the person and the box.

3. Is the frictional force on the person and the box the same? How does this allow the box to accelerate?

Problem-Solving Review

A 10.0 kg block resting on a 43.0° inclined plane is tied to a wall at the top of the incline with rope A that runs parallel to the plane. You pull on rope B that is attached to the other side of the block and also is parallel with the incline. Each rope can sustain a maximum tension of 100 N. You slowly increase the amount of force with which you pull on the rope.

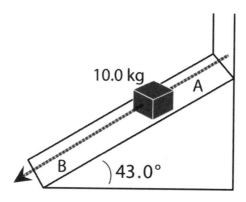

Set Up

1. Create a free-body diagram by drawing and labeling the forces acting on the block in the sketch above.

2. Which forces contribute to the tension in rope A? In rope B?

3. In which rope will the tension be greater? Which rope will break first?

Solve

1. What is the magnitude and direction of the force of gravity pulling down on the block?

2. Separate the force of gravity acting on the box into components in the directions parallel and perpendicular to the inclined plane. Label the parallel component the *x* component and the perpendicular component the *y* component. Add these components to your free-body diagram.

3. What are the magnitudes of the *x* and *y* components for gravity?

4. The block is at rest up until just before the rope breaks. What is the net force on the box in the *x* direction (down the incline)?

5. Write an expression for the *x* component of the net force acting on the block in terms of the *x* component of the force the instant before the rope snaps. (*Hint*: This is the sum of the forces in the direction parallel to the incline plane.)

6. Solve the expression you found in question 4 above for the force with which you are pulling on the block. What is the magnitude of this force when the tension in rope B is 100 N? How much force do you need to apply to break the rope?

Reflect

1. The moment before the rope breaks, what force is rope B exerting on the block. What force is rope B exerting on the wall?

2. What is the magnitude and direction of the block's acceleration immediately after the rope breaks if the tension in rope A is held constant?

Section 4-6 All problems involving forces can be solved using the same series of steps

Goal: Apply the sequence of steps used in solving all problems involving forces, including those that also involve kinematics.

Concept Check

1. You are standing in an elevator that is accelerating upward. What is the direction of the force the elevator is exerting on you? What about when the elevator is accelerating downward? Explain your answers.

2. When the elevator is accelerating upward, is its force on you greater, less than, or equal to the force of gravity on you? What about when the elevator is accelerating downward? Explain your answers.

Problem-Solving Review

An elevator in a fancy hotel has a 25.0 kg chandelier suspended by two ropes, each making a 20.0° angle with the vertical and connecting to the center of the chandelier. The elevator accelerates upward at 1.30 m/s².

Set Up

1. Sketch the chandelier and the ropes.

2. Create a free-body diagram by drawing and labeling the forces acting on the chandelier in your sketch above.

3. Write Newton's second law in component form for the chandelier.

Solve

1. What is the magnitude and direction of the net force acting on the chandelier?

2. What are the x and y components of the tension in each rope?

3. What is the magnitude of the tension in each rope?

Reflect

1. How fast would the elevator have to accelerate downward for the chandelier to appear weightless?

Section 4-7 Fish use Newton's third law and a combination of forces to move through water

Goal: Identify the forces that a fish uses for propulsion.

Problem-Solving Review
A fish weighing 10.0 kg can exert a force of 100 N with its caudal fin (that's the fin at the rear of the fish). If it moves its fin so that the force of the fish on the water is at 150° counterclockwise from its direction of motion, what is its acceleration in the forward direction? Assume that the water around the fish creates a constant drag force of 23.0 N while the fish is moving. (Note: This is an approximation. In reality the drag force will depend on the velocity of the fish.)

Set Up
1. Draw a free-body diagram of the fish indicating the force of the fish on the water and of the water on the fish. Also indicate the direction of the drag force. (*Hint*: The drag force opposes the fish's motion.)

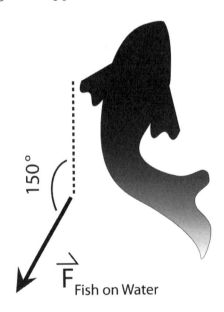

2. Draw the components of the fish's motion in the forward direction and in the lateral direction.

Solve

1. Write Newton's second law for the forces acting along the direction of motion. Take the lateral direction to be the *x* axis and the direction of motion to be the *y* axis.

2. What is the net force acting on the fish in the forward direction?

3. What is the magnitude of the fish's acceleration?

Reflect

1. What is the effect of gravity on the fish?

Chapter 5 Forces and Motion II: Applications

Section 5-1 We can use Newton's laws in situations beyond those we have already studied

Goal: Appreciate that Newton's laws apply to all forces and in all situations.

Concept Check

1. Name three types of forces and describe what causes them.

2. What are the physical causes of friction and drag?

Forces and Motion II

Section 5-2 The static friction force changes magnitude to offset other applied forces

Goal: Recognize what determines the magnitude of the static friction force.

Concept Check

1. For a block resting on a surface is there a friction force acting on it?

2. If the maximum force of static friction for an object is 15.0 N and you exert a 10.0 N force on the object, why doesn't it accelerate?

Problem-Solving Review

A block of mass 2.00 kg rests on a board with $\mu_s = 0.260$. The board is slowly lifted on one side until the block just starts to move.

Set Up

1. Draw a free-body diagram for the block before the board is lifted.

2. Draw a free-body diagram for the block after the board has been lifted so it makes a 10.0° angle with the horizontal. Be sure to label the components of the object's weight acting into the board and along the board's length.

Solve

1. Before the board is lifted, what is the maximum force that can be exerted on the block before it starts to slide?

2. Write the y components of $\sum F_{ext,y}$ for the block when the board is at $10.0°$. What is the normal force $\vec{n}$?

3. Write the x components of $\sum F_{ext,x}$ for the block when the board is at $10.0°$. What is the force of friction?

4. What is the maximum angle to which the board can be lifted before the block starts to slide?

Reflect

1. At $10.0°$ is the normal force on the block greater than, less than, or equal to the weight?

2. Is it possible for the coefficient of friction to be greater than one?

Section 5-3 The kinetic friction force on a sliding object has a constant magnitude

Goal: Be able to find the magnitude and direction of the force of kinetic friction.

Concept Check

1. You push a block up a ramp. It slides up the ramp, comes to a brief rest, and then slides back down. Draw velocity and friction vectors for the block on the way up and on the way down.

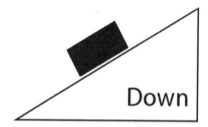
Down

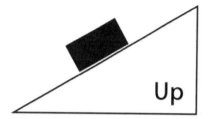
Up

2. How does the shape of the object change the force of friction?

Chapter 5

Problem-Solving Review

A block of mass 2.00 kg and a ball of mass 2.00 kg are released from the top of a 40.0° incline at height of 1.20 m from the ground. For the block $\mu_k =$ 0.340, and for the ball $\mu_r = 0.0600$.

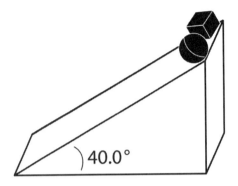

Set Up

1. Draw a free-body diagram for the block and the ball.

2. Find the normal force on the block and the ball.

Solve

1. Find the frictional force acting on the block and the ball as they roll down the incline.

2. What is the acceleration of the block on the incline?

3. What is the velocity of the block when it reaches the bottom of the incline?

4. The two objects encounter a flat surface at the bottom of the incline. What are the normal forces on the ball and the block on the flat surface?

5. What is the force of friction on each object on the flat surface (assume the same coefficients of friction as on the incline)?

Reflect

1. Explain whether the ball or the block goes further and why.

2. Does μ_k vary with velocity?

Section 5-4 Problems involving static and kinetic friction are like any other problem with forces

Goal: Solve problems in which the forces on an object include static or kinetic friction.

Concept Check
1. A slowly increasing force is exerted on an object at rest. In general, how does the magnitude of the frictional force change when the object starts to move, compared to right before it moves?

Problem-Solving Review
A 30.0 kg block is at rest on a ramp with an incline of 23.0°. A worker exerts a slowly increasing force on the box up the ramp. The worker has to apply 200 N before the box starts to move. The worker continues to push until the force on the box is 210 N.

Set Up
1. Draw a free-body diagram for the box just before it starts to move.

2. Draw a free-body diagram for the box when the maximum force is being exerted on it.

Solve

1. Write Newton's second law of motion for the block right before it starts to move.

2. What is the normal force on the block?

3. Find μ_s for the block right before it starts to move.

4. Write Newton's second law of motion for the block when the 210 N force is acting on it.

5. Find μ_k for the block. Assume the block accelerates up the incline at 0.950 m/s².

Reflect

1. What is the kinetic force of friction in this case?

Section 5-5 An object moving through air or water experiences a drag force

Goal: Analyze situations in which fluid resistance is important.

Concept Check
1. Explain why you can't use the equations involving velocity and acceleration learned in Chapter 2 when there is a velocity-dependent force acting.

2. What is the condition on the acceleration necessary for the ball to reach terminal velocity?

Problem-Solving Review
A penny with a mass of 2.50 g falls out of Axel's pocket from the top of the Empire State Building. The penny will reach terminal velocity long before it hits the ground.

Set Up
1. Draw a free-body diagram for the penny as it falls.

2. Write Newton's second law for the penny.

Solve

1. What is the force from the drag acting on the penny when it is traveling with a velocity of 0.200 m/s? Take the coefficient c to be 13.0 Ns²/m².

2. What is the acceleration of the penny at this instant?

2. What is the force from drag on the penny when it is at terminal velocity?

3. Find the terminal velocity of the penny.

Reflect

1. When the penny hits the ground, what force will the ground exert on the penny?

78   Chapter 5

Section 5-6 In uniform circular motion, the net force points toward the center of the circle

Goal: Apply Newton's laws to objects in uniform circular motion.

Concept Check

1. If you swing a rock on a string above your head level with the ground, which way is the rock accelerating?

2. A car is moving on a circular track. A passenger in the car feels a force toward the center of the track as the car moves. What is exerting that force on the passenger? What keeps the car from sliding off the track?

Problem-Solving Review

Aline is completing a physics lab for class. She takes a 0.300 kg block attached to a string and rotates the block at a constant speed. The block is attached to a string, and the string goes through a straw where it is connected to a 0.800 kg mass which is hanging at rest.

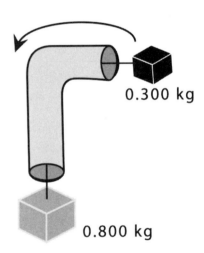

Set Up

1. Draw a free-body diagram for the 0.800 kg block.

2. Write Newton's second law for the 0.800 kg block.

3. Draw a free-body diagram for the 0.300 kg block. Neglect the force of gravity on the block. (This is a reasonable approximation because the force of tension is much larger than the force of gravity on the block.)

4. Write Newton's second law for the 0.300 kg block. Make the direction of the force toward the center of the circle.

Solve

1. Use Newton's second law to find the tension T on the 0.800 kg block.

2. What is the centripetal acceleration of the 0.300 kg block?

3. Use the tension T from question 1 above of this Solve section to find the centripetal acceleration of the 0.300 kg block.

4. The 0.300 kg block is 0.500 m away from the hole in the table at all times. What is its speed?

Reflect
1. If the string suddenly broke when the block was at an angle of $\theta = 0.00°$ with the horizontal while it was rotating counterclockwise, what would its velocity be?

Chapter 6 Work and Energy

Section 6-1 The ideas of work and energy are intimately related

Goal: Explain the relationship between work and energy.

Concept Check

1. What is the quantity that relates force to energy?

2. Name several types of energy.

82   Chapter 6

Section 6-2 The work that a constant force does on a moving object depends on the magnitude and direction of the force

Goal: Calculate the work done by a constant force on an object moving in a straight line.

Concept Check
1. Which quantities determine the amount of work being done on an object?

2. If you exert a force F while pushing a box a distance d, how much work does the box do on you?

Problem-Solving Review
You are pushing a 10.0 kg block with a horizontal force of 50.0 N along a surface. After a distance of 0.750 m there is an incline of 32.0° below the horizontal. The surface has a coefficient of kinetic friction of 0.360. You stop pushing the block after it lands on the incline.

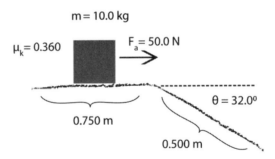

Set Up
1. Draw a free-body diagram of the block when it is on the surface.

2. Draw a free-body diagram of the block when it is on the incline.

Solve

1. What is the work done by you on the block when it is on the horizontal surface?

2. What is the work done by the block on you when it is on the horizontal surface?

3. What is the work done by friction on the block when it is on the horizontal surface?

4. What is the work done by friction on the block when it is on the incline if it moves a distance of 0.500 m along the surface of the incline?

5. What is the work done by gravity on the block?

6. What is the total work done on the block from all the forces acting on it from its starting position to 0.500 m down the incline?

Reflect

1. How much work is gravity doing on the block before it arrives at the incline?

Section 6-3 Kinetic energy and the work-energy theorem give us an alternative way to express Newton's second law

Goal: Describe what kinetic energy is and understand the work-energy theorem.

Concept Check

1. Describe in words how the work done on an object determines its kinetic energy.

2. Kinetic energy is a scalar quantity. Is work a vector or a scalar?

Problem-Solving Review

A tugboat pulls a container ship 1500 m, and as the tugboat moves a frictional force of 5000 N is exerted on it by the water. The tugboat's engine provides a force in the opposite direction of the force of friction. The tugboat is attached to the container ship by a taut wire cable, which forms an angle of 37.0° with the horizontal. The container ship experiences a force of friction of 7000 N.

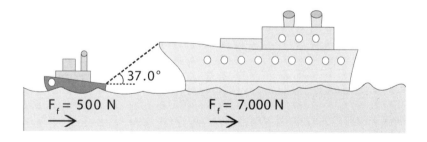

Set Up

1. What are the forces acting on the tugboat? Draw a free-body diagram. Split the tension in the cable into x and y components in your diagram.

2. What are the forces acting on the container ship? Draw a free-body diagram. Split the tension in the cable into x and y components in your diagram.

3. Does the y component of the tension in the cable contribute to the motion of the container ship?

Solve

1. If the maximum force the tugboat's engine can exert is 15000 N, what is the most work the tugboat can do on the container ship over 1500 m?

2. How much work is the container ship doing on the tugboat over that distance?

3. What is the change in kinetic energy of the container ship from the start to 1500 m?

4. If the container ship has a mass of 1.50×10^8 kg and an initial velocity of 0.00 m/s, what is the change in velocity of the container ship?

Reflect

1. How much work did the water perform on the container ship?

Section 6-4 The work-energy theorem can simplify many physics problems

Goal: Apply the work-energy theorem to solve problems.

Concept Check
1. You ride your bike up the same hill twice. The second time it takes you three times as long to get to the top. Did you do more work, less work, or the same amount of work the second time around?

Problem-Solving Review
A rocket of mass 8.00×10^4 kg takes off from rest traveling straight up. Its engine produces a constant thrust of 1.00×10^6 N. Use the work-energy theorem to calculate the height of the rocket when it has reached a speed of 25.0 m/s. You may ignore air resistance and changes in mass due to the consumption of fuel.

Set Up
1. Draw a free-body diagram of the forces acting on the rocket.

2. What is the initial kinetic energy of the rocket K_i? The final kinetic energy K_f?

Solve
1. How much work has the rocket done to reach a speed of 25.0 m/s?

2. Calculate the net force acting on the rocket.

3. What is the height of the rocket when it has reached a speed of 25.0 m/s?

Reflect
1. How would doubling the thrust of the rocket affect its speed at the height you calculated in question 3?

Section 6-5 The work-energy theorem is also valid for curved paths and varying forces

Goal: Recognize why the work-energy theorem applies even for curved paths and varying forces like the spring force.

Concept Check

1. You and your friend hike up a hill. You both have around the same weight. You run straight up the hill, but your friend strolls up a windy path. When you reach the top have you done more work, less work, or the same amount of work as your friend? Who has gained more potential energy?

2. A block is attached to the wall by a string. You pull the block so that it is just past its equilibrium point. Is the force on the block pointing toward you or toward the wall? Now you bring it back so that the spring is just shorter than the equilibrium position. In which direction is the spring force pointing now?

Problem-Solving Review

A block attached to a horizontal spring with spring constant $k = 10.0$ N/m on a frictionless surface is compressed to -0.500 m from its equilibrium position (at $x=0$) then released.

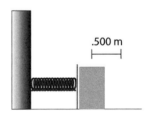

Set Up

1. What is the Force on the block when it is -0.500 m from equilibrium?

2. What is the total Force in the *x* direction on the block at 0.00 m?

3. What is the work done by gravity on the object?

Solve

1. Draw a Force vs. Position graph from *x* = −0.500 m to *x* = 0.00 m.

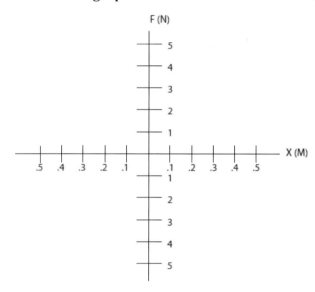

2. Use the graph you drew in part 1 to find the work done on the object as it moves from *x* = −0.500 m to *x* = 0.00 m. (Remember the equation $W = Fd$ cannot be used because it can only be applied to constant forces, and the spring force is dependent on position.)

3. What is the velocity of the object as it passes through the equilibrium position?

Reflect

1. Will the object's speed be zero or greater than zero at $x = 0.500$ m?

Now assume that when the block is at 0.250 m, a constant external force of −2.50 N is applied in the direction of motion until it reaches its equilibrium position.

Set Up

1. Draw a free-body diagram on the block at $x = 0.250$ m.

Solve

1. Draw a Force vs. Position graph from $x = 0.250$ m to $x = 0.00$ m.

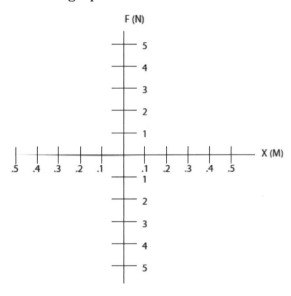

2. Use the graph you drew in the previous question to find the work done on the object as it moves from $x = 0.250$ m to $x = 0.00$ m.

Motion in Two or Three Dimensions

3. What is the velocity of the object as it passes through the equilibrium position?

Reflect
1. Will the object's acceleration be zero or greater than zero at $x = 0.00$m?

Chapter 6

Section 6-6 Potential energy is energy related to an object's position

Goal: Explain the meaning of potential energy and how conservative forces such as the gravitational force and the spring force give rise to gravitational potential energy and spring potential energy.

Concept Check

1. Explain how a stationary object can have the ability to do work, and give an example.

2. You raise a ball 2.00 m in the air and let it fall to the floor. Now you climb to the tenth floor of a building and raise a ball 2.00 m from the floor and let it fall. In which case is the ball's change in kinetic energy less than, greater than, or equal?

3. A mass is hanging on a spring attached to the ceiling. Write an expression for the total potential energy of the mass. Be sure to include gravitational as well as spring potential energy.

Problem-Solving Review

You pull a 10.0 kg bag of apples 15.0 m up a hill at a constant velocity and then let the bag slide down the hill. Assume that the hill is frictionless and the bag does not roll.

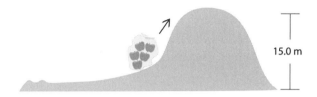

Set Up

1. Draw a free-body diagram of the forces acting on the bag as you pull it up the hill at a constant velocity.

2. Write an expression for the change in potential energy of the bag—from when it starts at the bottom of the hill to just before you release it at the top of the hill.

3. How is the change in potential energy of the bag as you bring it up the hill related to the change in kinetic energy as you let it slide down the hill?

4. Write an expression relating the kinetic energy of the bag to its instantaneous velocity.

Solve

1. What is the change in potential energy of the bag as you pull it to the top of the hill?

2. What is the change in potential energy of the bag as you let it slide back down?

3. What is the final velocity of the bag at the bottom of the hill?

Reflect

1. How much work did you do on the bag as you pulled it up the hill? How is the work you did related to the change in the bag's potential energy?

Section 6-7 If only conservative forces do work, total mechanical energy is conserved

Goal: Explain the differences between, and when you can apply, the generalized law of conservation of energy and the conservation of total mechanical energy.

Concept Check
1. List several non-conservative forces.

2. You push a chair across a room. As you first push on the chair, static friction prevents it from moving. Is static friction a conservative or non-conservative force? Does the force of static friction do work on the chair?

Problem-Solving Review
A mass of 3.00 kg is resting on a frictionless surface against a spring with spring constant $k=1000$ N/m. The spring is compressed by 0.250 m from equilibrium length and then released. The block slides on a frictionless level surface and then up an incline of $\theta = 30.0°$ until it comes to a brief rest.

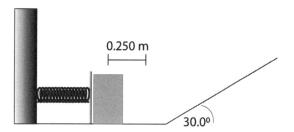

Set Up
1. What is the elastic potential energy U_e and gravitational potential energy U_{grav} before the block is released?

2. What is the elastic potential energy U_e and gravitational potential energy U_{grav} when the block is at its highest point?

3. What conservative and non-conservative work is being done on the block?

Solve

1. What is the gravitational potential energy of the block when it comes to rest on the ramp?

2. How much work was done by gravity on the block as it moved up the incline?

3. How far along the length of the incline does the block go?

Reflect

1. Is total mechanical energy conserved?

Now consider the same situation, but assume the surface and the ramp have a nonzero coefficient of friction. After being launched by the spring, the block slides a distance of 1.50 m up along the ramp.

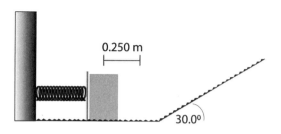

Set Up
1. What are the conservative and non-conservative forces acting on the block?

2. Is the work done by friction positive or negative?

3. What is the total mechanical energy right as the block is released?

Solve
1. If the block travels 1.50 m along the incline, what was its maximum height?

2. What work was done by non-conservative forces?

Reflect

1. What is the total mechanical energy when the block comes to its maximum height on the incline?

2. Was mechanical energy conserved? Was total energy conserved?

3. Where did this energy go?

Section 6-8 Potential energy is energy related to an object's position

Goal: Identify which kinds of problems are best solved with energy conservation and the steps to follow in solving these problems.

Concept Check

1. Assume you are asked to solve for the velocity of a swinging pendulum. What are the advantages of using conservation of energy over just considering the forces?

Problem-Solving Review

A 2.00 kg block slides down a curved ramp. If the block starts at a height of 10.0 m, what is its speed when it reaches the bottom of the ramp? Assume the bottom of the ramp is 2.00 m off the ground and that the block loses 50.0 joules of energy to friction.

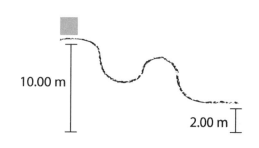

Set Up

1. Draw a free-body diagram of the forces acting on the block.

2. Describe sources that contribute to the potential energy of the block.

3. What is the initial kinetic energy of the block K_i at the top of the ramp?

Solve

1. Write an expression for the total energy of the block at the top of the ramp and at the bottom of the ramp. Write your two expressions in terms of the potential energy at the top and the bottom (PE_1 and PE_2), the kinetic energy at the top and bottom (KE_1 and KE_2), and the energy lost to friction (W_f). How are they related? What is the difference between the final mechanical energy and the final total energy?

2. Calculate the total energy of the block at the top of the ramp.

3. Equate the two expressions you found in question 1 and solve for the final kinetic energy K_f of the block.

4. Calculate the speed of the block at the bottom of the ramp.

Reflect

1. For a frictionless ramp would the block move faster if its mass were only 1.00 kg or with its original mass of 2.00 kg?

Chapter 7 Momentum, Collisions, and the Center of Mass

Section 7-1 Newton's third law helps lead us to the idea of momentum

Goal: Comprehend the significance of momentum and the center of mass.

Concept Check

1. What two physical quantities compose momentum?

2. Which of Newton's laws deals with two bodies interacting? Write the equation for this law.

Section 7-2 Momentum is a vector that depends on an object's mass, speed, and direction of motion

Goal: Define the linear momentum of an object and explain how it differs from kinetic energy.

Concept Check

1. You are pushing a large moving box that is twice your weight down the street at a constant velocity. Which has greater momentum, you or the box? Explain your answer.

2. Is the direction of the box's momentum the same or opposite of your momentum? Explain your answer.

Problem-Solving Review

A 30,000 kg space shuttle starting at rest on Earth is launched and reaches a velocity of 28,000 km/hr. Assuming the shuttle reached this speed very quickly, how much momentum did Earth acquire from the shuttle pushing off? What is the change in velocity of Earth as a result? (The mass of Earth is approximately 5.97×10^{24} kg.)

Set Up

1. On the diagram draw the directions of shuttle's and Earth's momentum vectors, $\vec{p}_{Shuttle}$, $\vec{p}_{Earth}$.

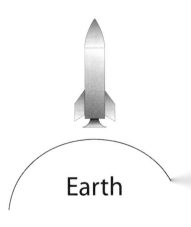

2. Is Earth's momentum greater than, less than, or equal to the shuttle's?

Solve

1. What is the magnitude of the shuttle's momentum? What is the magnitude of Earth's momentum?

2. Write an expression for Earth's velocity in terms of the magnitude of the Earth's momentum and its mass.

3. Solve the expression you found in question 2 above for Earth's velocity. What is the direction of the Earth's velocity?

Reflect

1. Write an expression for the magnitude of Earth's kinetic energy in terms of its mass and velocity acquired from the space shuttle's launch. Is Earth's kinetic energy greater than, less than, or equal to that of the space shuttle?

2. Solve the expression you found in the previous question for the kinetic energy of Earth.

Section 7-3 The total momentum of a system of objects is conserved under certain conditions

Goal: Explain the conditions under which the total momentum of a system is conserved, and why total momentum is conserved in a collision.

Concept Check
1. Two objects collide in deep space. List the internal and external forces.

2. Two objects are sliding on a table and collide. List the internal and external forces.

3. Explain why in both cases described above external forces can be ignored.

Problem-Solving Review
Object a, which has a mass of 6.20 kg, is sliding on the +x axis at 1.00 m/s on a frictionless surface when object b, with a mass of 2.50 kg moving at 3.00 m/s in the +x axis direction, comes up from behind and collides with it. After the collision, object a is observed to move in direction $\theta_{a,f} = 3.00°$ below the +x axis and $\theta_{b,f} = 7.00°$ above the +x axis.

Set Up
1. Draw a diagram of the situation before and after the collision. Label the velocities, masses of the objects, and the angles they move off at.

2. Are there any net external forces acting on the system? Will momentum be conserved?

3. Write $\vec{p}_{a,ix}$ and $\vec{p}_{b,ix}$, the x component of the initial momentum for a and b respectively. Find $\vec{p}_{ix}$ in terms of $\vec{p}_{a,ix}$ and $\vec{p}_{b,ix}$.

4. Write $\vec{p}_{a,iy}$ and $\vec{p}_{b,iy}$.

5. Write $\vec{p}_{a,fx}$ and $\vec{p}_{b,fx}$ (the final momentum in the x direction of object a and object b) in terms of quantities given in the problem and $\vec{v}_{a,f}$ or $\vec{v}_{b,f}$ respectively. Find $\vec{p}_{fx}$ in terms of $\vec{p}_{a,fx}$ and $\vec{p}_{b,fx}$.

6. Write $\vec{p}_{a,fy}$ and $\vec{p}_{b,fy}$ in terms of quantities given in the problem and $\vec{v}_{a,f}$ or $\vec{v}_{b,f}$ respectively. Find $\vec{p}_{fy}$ in terms of $\vec{p}_{a,fy}$ and $\vec{p}_{b,fy}$.

Solve

1. Use conservation of momentum to find the relation between $\vec{p}_{ix}$ and $\vec{p}_{fx}$. What are the two unknowns in this equation?

2. Solve the equation in the previous question for $\vec{v}_{a,f}$. You should have an expression for $\vec{v}_{a,f}$ in terms of $\vec{v}_{b,f}$ and known quantities.

3. Plug your expression for $\vec{v}_{a,f}$ into the conservation of momentum equation you found for $\vec{p}_{ix}$ and $\vec{p}_{fx}$. Now solve for the only unknown, $\vec{v}_{b,f}$.

4. Find $\vec{v}_{a,f}$.

Reflect

1. Find the total momentum before and after the collision. Are they equal?

Section 7-4 In an inelastic collision, some of the mechanical energy is lost

Goal: Identify the differences and similarities between elastic, inelastic, and completely inelastic collisions

Concept Check

1. For each of the following collisions, state whether the collision is elastic, inelastic, or completely inelastic and explain your answer.

    a. A bouncy ball hits the ground, deforms, and then bounces back upward.

    b. You hit a mosquito with a fly swatter.

    c. Two billiard balls collide.

2. Mark is driving and a piece of paper hits his windshield and sticks there while he continues to drive. Is this collision elastic, inelastic, or completely inelastic? Is mechanical energy conserved in this collision?

Problem-Solving Review

A 10.0 kg meteorite collides with Earth at 30.0 km/s forming a large crater. How much mechanical energy is lost in this collision? (The mass of Earth is approximately 5.97×10^{24} kg.)

Set Up
1. Is this collision elastic, inelastic, or completely inelastic? Explain.

2. What is the magnitude of the meteorite's momentum before it collides with Earth? (Assume Earth's velocity is zero before the collision.)

Solve
1. Write an expression for conservation of momentum during the collision of the meteorite with Earth.

2. Solve the expression you found in the previous problem for the final velocity of the combined object formed by the meteorite stuck to Earth.

3. What is the initial kinetic energy of the system? What is the final kinetic energy of the system?

4. How much mechanical energy was lost in the system?

Reflect
1. If this had been an elastic collision and the meteorite bounced off Earth without deforming, what would its final velocity be?

Section 7-5 In an elastic collision, both momentum and mechanical energy are conserved

Goal: Apply momentum conservation and energy conservation to problems about elastic collisions.

Concept Check

1. Write the conservation equations that can be used in elastic collisions.

2. Write the conservation equations that can be used in inelastic collisions.

Problem-Solving Review

An object with a mass of 1.00 kg flies through space at 15.0 m/s in the +x direction. It catches up to and collides with another object of mass 0.120 kg moving at 10.0 m/s in the same direction.

Set Up

1. Write the conservation of momentum equation for the system.

2. Write the conservation of energy equation for the system.

Solve

1. Solve the conservation of momentum equation for the final velocity of the second object, $\vec{v}_{2,f}$, in terms of the final velocity of the first object, $\vec{v}_{1,f}$.

2. Now consider the conservation of energy equation you found in Set Up question 2. Substitute the relation for $\vec{v}_{2,f}$ you found in the last question into the conservation of energy equation. After completing this step, the only unknown is $\vec{v}_{1,f}$. Write this equation as a quadratic equation, $ax^2+bx+c=0$.

3. Solve the quadratic equation you found in the last question. This will give $\vec{v}_{1,f}$. What is the value of $\vec{v}_{1,f}$?

4. What is $\vec{v}_{2,f}$?

Reflect

1. How would the direction of the second object change after the collision for the case $m_2 > m_1$?

Momentum, Collisions, and the Center of Mass

Section 7-6 What happens in a collision is related to the time the colliding objects are in contact

Goal: Relate the momentum change of an object, the force that causes the change, and the time over which the force acts.

Concept Check
1. Give two examples of units that can be used for impulse.

Problem-Solving Review
A crash test dummy is being used to test airbags in a newly designed car. The car has a velocity of 150 km/hr before it collides with a wall. Let's calculate how long the crash test dummy should be acted on by the airbag in order to reduce the average force on the dummy to 100 N.

Set Up
1. In the sketch below draw vectors indicating (a) the force of the air bag on the crash dummy, (b) the momentum of the crash test dummy, and (c) the momentum of the car during the collision.

2. Calculate the momentum of the car just before the collision. What is the momentum of the car at the end of the collision?

Solve
1. What is the change in the crash test dummy's momentum due to its interaction with the airbag?

2. Write an expression for the impulse on the crash test dummy in terms of the change in its momentum.

3. Use the expression you found in the previous question to find an expression for the time the crash test dummy interacts with the air bag.

4. How long should the crash test dummy be acted on by the airbag in order to reduce the average force on the dummy to 100 N?

Reflect

1. Before the collision the crash test dummy was traveling at the same velocity as the car. During the collision the crash test dummy decelerates as it is being stopped by the air bag. What is the average acceleration of the crash test dummy during the time it interacts with the airbag?

2. Is this acceleration stronger or weaker than the acceleration due to gravity near the surface of Earth?

Section 7-7 The center of mass of a system moves as though all of the system's mass were concentrated there

Goal: Explain the physical significance of the center of mass, and describe how the net force on a system affects the motion of the system's center of mass.

Concept Check
1. If a firework explodes in the air, how does this affect the center of mass?

Problem-Solving Review
It is possible to make a humane mouse trap by taking a cardboard tube, putting a tiny bit of peanut butter on one end and then setting it on a counter with one end sticking out over the edge. When the mouse runs to get the peanut butter, the tube and the mouse fall into a bucket waiting on the ground. The tube is 0.300 m long and has a mass of 0.010 kg. It sits so that 40.0% of its length is out over the edge. The mouse has a mass of 0.020 kg.

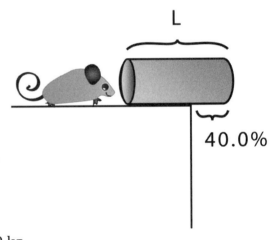

Set Up
1. Assume that the mouse runs into the tube toward the peanut butter. Describe the condition for the mouse and tube to fall off the counter? Assume that the origin is at the left-hand side of the tube.

2. Right when the center of mass of the mouse is at the origin, what is the center of mass x_{CM} for the mouse and paper towel tube system? What about when it has moved a distance of 0.150 m?

Solve

1. If the mouse runs at a velocity of 2.00 m/s and is toward the end of the tube and is at 0.100 m, what is the total momentum of the system?

2. What is the position of the mouse when the mouse and paper towel tube system fall off the counter?

3. Assume that when the system falls, the mouse holds onto the walls of the tube. What is the external force on the system as it falls?

Reflect

1. If the mouse tries to run up the tube as they both fall, will this change the center of mass of the system?

Chapter 8 Rotational Motion

Section 8-1 Rotation is an important and ubiquitous kind of motion

Goal: Define translation and rotation.

Concept Check

1. In the following situations describe whether the object is undergoing a translation, rotation, or both. Explain your answer.

a. A barrel is rolling down a hill.
- a) translation
- b) rotation
- c) both

b. An elevator car traveling upward.
- a) translation
- b) rotation
- c) both

c. A CD spinning in a computer.
- a) translation
- b) rotation
- c) both

Section 8-2 An object's rotational kinetic energy is related to its angular velocity and how its mass is distributed

Goal: Explain what is meant by the moment of inertia of an object and how to use it to calculate the rotational kinetic energy of a rotating object.

Concept Check

1. A turntable with radius r rotates. Is the angular velocity at a point $.r$ from the axis of rotation smaller, larger, or the same compared to a point a distance r from the axis of rotation? How about the translational velocity?

2. Can two objects with the same mass have different moments of inertia? Why or why not?

Problem-Solving Review

A turntable rotates at a constant angular velocity $\omega=0.500$ rad/s. A small stone of mass 0.0130 kg sits on the turntable a distance 0.0500 m from the center.

Set Up

1. Convert 0.500 rad/s to rev/min.

2. What is the speed of the stone?

3. What is the moment of inertia for the stone?

Solve

1. Find the rotational kinetic energy of the stone.

2. Find the kinetic energy of the stone.

Reflect

1. How long will it take for the stone to go around 10 times? Would this change if the stone was moved closer to the center?

Section 8-3 An object's moment of inertia depends on its mass distribution and the choice of rotation axis

Goal: Describe the techniques for finding the moment of inertia of an object, including use of the parallel-axis theorem.

Concept Check

1. A pro basketball player spins a basketball on his finger. If the ball were to be inflated more, increasing its radius, would its moment of inertia be smaller, larger, or stay the same? Explain your answer.

2. Is a flying disc's moment of inertia larger if you spin it around its center or around its rim? Explain your response.

Problem-Solving Review

Kaylie swings a 0.500 kg yoyo around in a circle by the end of its 0.400 m string at 6.30 rad/s. The yoyo has the shape of a solid cylinder with radius 3.00 cm and thickness 6.00 cm. We will use the parallel-axis theorem to find the moment of inertia of the yoyo and subsequently its rotational kinetic energy.

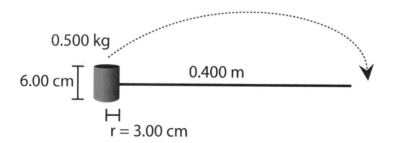

Set Up

1. On the sketch of the yoyo mark the center of mass of the yoyo.

2. On the sketch, draw where the axis of rotation would be if the yoyo were rotating about its center of mass.

Solve

1. What is the moment of inertia for a solid cylinder, as described above, rotating about its center of mass?

2. Using the parallel-axis theorem, write an expression for the moment of inertia of the yoyo when it's being swung around in a circle by the end of its string. Your expression should be in terms of the moment of inertia you found in the previous question, the mass of the yoyo, and the length of the string.

3. What is the moment of inertia of the yoyo when it's being swung around in a circle by the end of its string?

4. Write an expression for the yoyo's rotational kinetic energy.

5. Solve the expression you found in the previous question for the yoyo's rotational kinetic energy.

Reflect

1. What is the moment of inertia for an identical yoyo in the same situation, but with a mass of 0.300 kg?

2. If the yoyo's length were to be decreased, would its moment of inertia be smaller, larger, or stay the same?

Chapter 8

Section 8-4 Conservation of mechanical energy also applies to rotating objects

Goal: Apply the conservation of mechanical energy to rotating objects, including objects that roll without slipping.

Concept Check
1. Write the equation for the total mechanical energy of a ball rolling.

Problem-Solving Review
A block and a ball each of 3.00 kg are moving side by side, each with total kinetic energy of 50.0 J on a flat frictionless surface. (The ball has a radius of 0.100 m and is rolling without slipping.) After a short while, they encounter an incline of $\theta = 20.0°$.

Set Up
1. Write the conservation of energy equation for the block in terms of $K_{translational}$ and U. Use the bottom of the incline as the initial position and the maximum height it reaches on the incline as the final position.

2. Write the conservation of energy equation for the ball in terms of $K_{translational}$, $K_{rotational}$, and U. Use the bottom of the incline as the initial position and the maximum height it reaches on the incline as the final position.

3. Write an expression for the potential energy U of the block and another expression for the ball.

4. Find the moment of inertia of the ball.

5. Write an expression for the total kinetic energy of the ball only in terms of its mass, radius, its moment of inertia, and its translational velocity.

Solve
1. What is the translational velocity of the block? What is the maximum vertical height it will reach on the incline?

2. What is the translational velocity of the ball?

3. What is the maximum vertical height the ball will reach on the incline?

Reflect
1. If the ball were to slide without rolling, would its translational velocity be larger, smaller, or stay the same?

Section 8-5 The equations for rotational kinematics are almost identical to those for linear motion

Goal: Identify the equations of rotational kinematics and know how to use them to solve problems.

Concept Check
1. The speed of a car is increasing, and the wheels are rotating in the clockwise direction. Is the angular velocity of the front driver's side wheel positive or negative? Is its angular acceleration positive or negative?

Problem-Solving Review
On a hot summer day John turns on a fan to cool off. The fan moves three quarters of the way around in 0.500 seconds. Let's calculate the fan's angular acceleration and final angular velocity.

Set Up
1. Assume the initial angular position is 0. What is the final angular position of the fan in radians?

2. Fill out the following Know/Don't Know table. Fill in the value of the quantity if you know it (some quantities maybe zero). If you need to calculate the value of the quantity to solve the problem, fill in the box with a question mark. If the equation doesn't use the quantity just fill in a dash. Some quantities have been filled in for you.

Equation	Quantity					
	t	θ	θ_0	ω	ω_0	α
$\omega = \omega_0 + \alpha t$	0.5s	–			–	
$\theta = \theta_0 + \omega_0 t + \frac{1}{2}\alpha t^2$	0.5s		0	–		

Solve

1. Which equation should be used to find the fan's angular acceleration? Rearrange this equation, solving for the angular acceleration in terms of the other quantities.

2. Use the equation you found with your values from the Know/Don't Know table to find the fan's angular acceleration.

3. Which equation should be used to find the fan's final angular velocity? Rearrange this equation, solving for the final angular velocity in terms of the other quantities.

4. Use the equation you found with your values from the Know/Don't Know table to find the fan's final angular velocity.

Reflect

1. Convert your answer in question 4 of the Solve section to rev/s.

Chapter 8

Section 8-6 Torque is to rotation as force is to translation

Goal: Define the concept of lever arm and know how to use it to calculate the torque generated by a force.

Concept Check
1. Explain in words why door knobs are located far away from the hinge.

2. What quantity has to be non-zero if the angular velocity of a rotating object changes? Explain your answer.

Problem-Solving Review
A pottery wheel of mass 5.00 kg and radius 0.300 m is spinning with an angular velocity of 20.0 rad/s. The motor exerts a torque of 5.00 N·m in the clockwise direction.

Set Up
1. The pottery wheel is shaped like a flat disc. Find the moment of inertia of the disc about its center. (*Hint*: You can use the formula for the moment of inertia of a solid cylinder and set its length l to 0.)

2. Write Newton's second law for rotation for the pottery wheel.

Solve
1. What is the angular acceleration of the wheel?

2. You exert a force of 7.00 N at the edge of the wheel in the counterclockwise direction. How much torque are you exerting on the wheel? What is the net torque on the wheel?

3. What is the new angular acceleration of the wheel?

4. Now you exert a force of 7.00 N in the counterclockwise direction 0.150 m from the axis of rotation. How much torque are you exerting on the wheel? What is the net torque on the wheel?

5. What is the angular acceleration of the wheel?

Reflect

1. What is the angle between $\vec{r}$ (the vector between the center of rotation and where the force is applied) and the direction of the counterclockwise force?

130   Chapter 8

Section 8-7 The techniques used for solving problems with Newton's second law also apply to rotation problems

Goal: Explain the meaning of Newton's second law for rotational motion.

Concept Check

1. You are attaching a shelf to a wall with a hand-driven screwdriver. Is it easier to turn screws if the handle of the screwdriver is thick or thin? Explain your response.

2. Stefan is spinning Mark around in an office chair. Assuming Stefan always pushes with the same amount of force, is it easier to spin Mark if he holds his arms out or closer to his body? Explain your answer.

Problem-Solving Review

A 30.0 kg block is attached to a 50.0 kg block by a rope that is looped around a pulley. We will find the acceleration of both blocks and the angular acceleration of the pulley. Assume the pulley is a uniform disk with a mass of 2.00 kg and a radius of 1.25 cm.

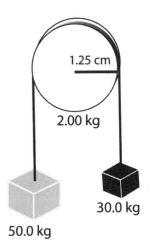

Set Up

1. Draw a free-body diagram indicating the forces acting on each block and the pulley.

2. If you assume that the rope doesn't stretch, what can you say about the relation between the velocities of the blocks and the velocity of the pulley at its rim? Write this relationship down as an expression.

3. Write an expression for the moment of inertia of the pulley in terms of its mass and radius.

Solve

1. Write Newton's second law for the rotational motion of the pulley.

2. Write Newton's second law for each block.

3. Find the net force $\sum F_{ext,y}$ for each block. Plug this net force into the statements of Newton's second law you found in the previous question to find the equations of motion for each block. (*Hint*: Each block may have a different tension force pulling upon it.)

4. Find the net torque $\sum \tau_z$ acting on the pulley. Apply this net torque to the final statement of Newton's second law you found to find the equation of motion for the pulley.

5. Use the expression you found in question 2 of the Set Up to relate the accelerations of each block to the angular acceleration of the pulley. (You should find two equations.)

6. These two equations plus the three equations of motion you found provide you with five equations with five unknowns. Combine these equations to find expressions for the angular acceleration of the pulley and the linear accelerations of both blocks.

7. Solve these equations for the angular acceleration of the pulley and the linear accelerations of both blocks.

Reflect

1. Find the angular velocity of the pulley after one second.

2. What would the moment of inertia of the pulley be if its axis of rotation were at its rim instead of its center?

Section 8-8 Angular momentum is conserved when there is zero net torque on a system

Goal: Define what is meant by the angular momentum of a rotating object and of a moving particle, and explain the circumstances under which angular momentum is conserved.

Concept Check

1. In college one of the authors was picked to do a demonstration where he stood on a pedestal that was free to rotate. He held a 1.00 kg mass in each hand with his arms fully extended while the professor gave him a push. He spun slowly. After a bit he retracted his arms and began to spin much more quickly. Did the moment of inertia increase, decrease, or stay the same when the masses were brought close to the author? Explain your answer.

2. For the case above, explain whether the angular momentum increases, decreases, or stays the same.

Problem-Solving Review

An asteroid of mass 3.00×10^4 kg and velocity 2.50×10^4 m/s is heading toward Earth. It is roughly 4.00×10^{10} m away from Earth. Traveling in a straight line, it should miss the center by 2.00×10^6 km.

Set Up

1. We will call the vector $\vec{r}$ the distance from the center of rotation of Earth to the asteroid. Draw a diagram of the situation and label the vector $\vec{r}$.

2. Label $\vec{p}$, the momentum vector for the asteroid.

3. Label $\vec{R}$ the vector from the axis of rotation to the nearest path of the particle. $\vec{R}$ is perpendicular to the path of the particle.

Solve

1. Find $\sin \phi$, where ϕ is the angle between $\vec{r}$ and $\vec{p}$.

2. What is the component of the asteroid's momentum perpendicular to $\vec{r}$?

3. Find the angular momentum for the asteroid about Earth.

Reflect

1. What is the asteroid's moment of inertia about Earth?

2. What is the asteroid's angular velocity?

Section 8-9 Rotational quantities such as angular momentum and torque are actually vectors

Goal: Explain how to find the direction of the angular momentum, angular velocity, and torque vectors.

Concept Check

1. Ben is riding his bicycle forward. Is the angular momentum vector of his front wheel pointing to his left or his right? Explain your answer.

Problem-Solving Review

Blythe is bench pressing weights at the gym. The bench has two supports separated by 1.10 m that hold a 1.75 m long bar which is to be pressed. The bar is 22.5 kg and sits evenly across the supports. Blythe starts to add weights to the right side of the bar. We will calculate how much mass Blythe can add to just the right side of the bar before it tips over. (Assume that the weights sit at 0.20 m from the end of the bar.)

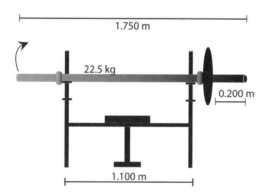

Set Up

1. You can assume that the force of gravity acting on the bar itself (not including the added weights) acts as the bar's center of mass. Where is the bar's center of mass?

2. Draw a free-body diagram including all of the forces acting on the bar.

3. Where is the axis of rotation around which the bar will rotate if it tips over?

4. What will the net torque on the bar be just before it tips over?

Solve

1. Write an expression for the net torque on the bar in terms of the torque due to the center of mass of the bar τ_{CM}, the torque due to the weights τ_w, and the net torque you found in the previous question.

2. What is the torque on the bar due to its own mass? What is its direction?

3. What is the torque on the bar due to the weights added by Blythe in terms of the mass m that she adds?

4. Solve your expression for the net torque to find how large of a mass can be added to the bar before it tips over.

Reflect

1. If the bar starts to tip over, in which direction will its angular momentum vector point?

Chapter 9 Elastic Properties of Matter: Stress and Strain

Section 9-1 When an object is under stress, it deforms.

Goal: Explain the meaning of stress and strain in physics.

Concept Check
1. Describe the difference between stress and strain. Will the units of stress and strain be the same or different?

2. Give three examples of elastic objects and three examples of plastic objects (not plastic in the everyday sense of the word, but an object made of a material that doesn't return to its original shape when deformed).

Section 9-2 An object changes length when under tensile or compressive stress

Goal: Define Hooke's law for an object under tension or compression.

Concept Check

1. Two springs are made of aluminum. One spring is twice as thick as the other. Will they have the same Young's modulus, spring constant, or both?

2. Explain the difference between compressive stress and compressive strain. What are the units of compressive strain?

Problem-Solving Review

A 10.0 kg mass is suspended vertically by a steel band. The rest length of the band is 0.330 m and has a radius of 20.0 cm. We will calculate the change in length of the steel band due to the mass that is hanging on it. (Ignore the effect of the spring's own weight.) Young's modulus for steel is 1.80×10^{10} N/m².

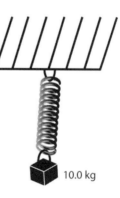

10.0 kg

Set Up

1. Draw a free-body diagram of all of the forces acting on the mass.

2. What is the cross-sectional area of the band?

3. What is the magnitude of the force pulling down on the block?

Solve
1. What is the stress on the steel band?

2. Find the spring constant k.

3. Use Hooke's law to find an expression for the change in length of the band.

Reflect
1. What is the strain in the material?

Elastic Properties of Matter: Stress and Strain

Section 9-3 Solving stress–strain problems: Tension and compression

Goal: Solve problems that involve the relationships between tensile stress and strain and between compressive stress and strain.

Concept Check
1. What is the difference between tensile stress and strain and compressive stress and strain?

2. How is the spring constant k related to Y, Young's modulus?

Problem-Solving Review
In physics lab, Li takes a steel cube (Young's modulus 1.80×10^{10} N/m²) with sides of length 0.0500 m and exerts a compressive force of 1000 N on it. Then he pulls on the cube with a force of 1000 N while it is attached to the wall.

Set Up
1. What is the area of each side of the cube?

Solve
1. What is the tensile stress on the cube? What is the compressive stress on the cube?

2. What is the change in length from the applied force when the object is being compressed?

3. What is the compressive strain on the cube? What is the tensile strain on the cube?

Reflect

1. What is the spring constant k of the cube?

2. If the cube were to be made from a material with a larger Young's modulus, how would the change affect the change in length of the cube?

Section 9-4 An object expands or shrinks when under volume stress

Goal: Explain the importance of the minus sign in the relationship between volume stress and volume strain.

Concept Check

1. What are the units of volume strain?

2. Imagine a cube and a sphere each made out of steel and each occupying the same volume. An equal amount of force is applied to the entire surface of each object. Which object will exhibit a greater change in volume? Explain your answer.

Problem-Solving Review

A cube-shaped trash compactor is filled with glass materials. (Assume the glass occupies the entire volume of the trash compactor.) The compactor must increase the pressure on the glass by 4.00×10^7 Pa in order to change its volume by 0.100%. We will find the bulk modulus of the glass.

Set Up

1. Is this an instance of compressive or tensile volume stress? Explain.

2. What is the volume strain of the glass?

Solve

1. Find an expression for the bulk modulus of the glass in terms of the change in pressure Δp (simplify using your answer from question 2 in the Set Up).

2. Find the bulk modulus of the glass.

Reflect

1. What percent of the original volume of the glass would the cube of glass occupy if the pressure were doubled?

Section 9-5 Solving stress–strain problems: Volume stress

Goal: Solve problems in which an object is under volume stress.

Concept Check

1. If an object goes from a higher pressure region to a lower pressure region, how does this affect the volume strain?

2. If an object goes from a lower pressure region to a higher pressure region how does this affect the volume strain?

Problem-Solving Review

Consider again the steel cube Zhu Song was experimenting on in Section 9-3. Each side of this cube has length of 5.00 cm. The bulk modulus, B, is 1.60×10^{10} N/m². At Earth's surface, atmospheric pressure is 1.00×10^{5} N/m². To continue his investigations into the properties of the steel cube, Zhu Song brings the cube into space where the pressure is zero.

Set Up

1. Find the volume of the cube.

2. What is the force exerted on each side of the cube by the atmospheric pressure?

Solve

1. What is the volume stress on the cube as Zhu Song brings it from the surface of Earth into space?

2. What is the change in volume of the cube from when it is on the surface of Earth to when it is in space?

3. What is the volume strain on the cube?

Reflect

1. Take the volume strain, $\frac{\Delta V}{V_0}$ and multiply it by $-B$. To what quantity is this equal?

2. What is the change in side length between the cube being on the surface and being in space?

Elastic Properties of Matter: Stress and Strain

Section 9-6 A solid object changes shape when under shear stress

Goal: Define what is meant by shear, and explain what shear stress and shear strain are.

Concept Check
1. What is shear stress?

2. A shear force on a solid causes it to deform. What is the effect of a shear force on a liquid or a gas?

Problem-Solving Review
A rubber block with a length of 23.0 cm, width of 17.0 cm, and a height of 21.0 cm rests on a surface. You exert a 500 N force on the top of the block at an angle 15.0° below the horizontal. Upon exerting the force, you see the block deform by about -4.20×10^{-3} m.

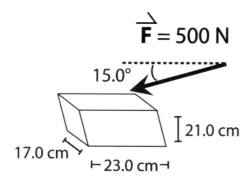

Set Up
1. What is the surface area over which the force is acting?

148 Chapter 9

2. What is $F_\parallel$, the component of the force acting parallel to the surface of the block?

Solve

1. What is the height of the block after the force is applied?

2. What is the shear stress on the block?

3. What is the shear strain?

4. What is the shear modulus for rubber?

Reflect

1. The force you exerted was in one direction, but shear requires two forces acting in opposite directions. For this case, what provided the other force?

2. What assumption was made about the magnitude of this force?

Section 9-7 Solving stress–strain problems: Shear stress

Goal: Solve problems that involve the relationship between shear stress and shear strain.

Concept Check

1. If the shear stress on an object is increased, does the shear strain go up, down, or stay the same? Explain your answer.

2. If the surface area of an object that is being sheared increases, how does this affect the shear stress? Explain your answer.

Problem-Solving Review

Plastic bags are commonly made of polyethylene, which has a shear modulus of 1.20×10^8 N/m², and are around 2.50×10^{-5} m thick. Consider a 0.500 m² sheet of this material with a 15.0 N force acting in opposing parallel directions on each side.

Set Up

1. Sketch the sheet with the appropriate forces acting on it.

2. Label x and h on your drawing.

150 Chapter 9

Solve

1. Find the shear stress on the sheet.

2. Find the shear strain on the sheet.

3. Upon shearing, there will be a deformation in the sheet. What is the angle of this deformation with the horizontal?

Reflect

1. If everything in the problem was the same, but the material was steel with a shear modulus of 7.90×10^{10} N/m², how would the magnitude of the strain change?

Section 9-8 Objects deform permanently or fail when placed under too much stress

Goal: Explain what must happen to a material for it to go from elastic to plastic to failure, and why biological materials do not become plastic.

Concept Check

1. What properties of a material change when it goes from an elastic regime to a plastic one?

2. What happens to a material when it fails?

Problem-Solving Review

In a physics experiment you are stretching a spring and obtain the following results:

Applied Force (N)	100	200	300	400	500	600
Stretch of Spring (m)	.060	.120	.180	.240	.250	.250

Set Up

1. Sketch a rough plot of the F vs. x for the table on the other page.

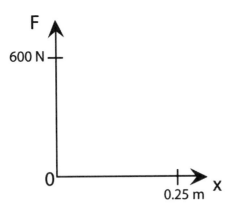

2. In the sketch above label the elastic and the plastic region.

Solve

1. From the data, find the region where the spring is elastic.

2. Find the region in the data where the spring is plastic.

3. Calculate k for the spring.

4. Take the plastic region you found above and assume the unstretched length of the spring is 0.260 m; find the strain on the spring.

Reflect

1. Will the spring return to its un-stretched size of 0.260 m after the 600 N force ceases to be exerted on it?

2. What would you guess would happen to the spring if more weight was placed on it?

Section 9-9 Solving stress–strain problems: From elastic behavior to failure

Goal: Solve problems concerning the stresses and strains on materials that may not obey Hooke's law.

Concept Check

1. What does it mean mathematically when an object doesn't obey Hooke's law?

2. In which regime do biological materials not obey Hooke's law?

Problem-Solving Review

The web of a Darwin's Bark Spider has an average breaking stress of around 1.80×10^{11} N/m², and an average strain of 0.330.

Set Up

1. If the spider silk of a Darwin's Bark Spider with area A=1.00×10^{-6} m² is at breaking stress, what is the maximum force acting on the strand?

Solve

1. Find Young's modulus for the spider silk at the breaking point.

2. If the piece of silk is initially 0.500 m long, how much will its length change when the maximum force is acting on it?

Reflect

1. What would happen to the spider silk in this case if you stopped exerting the applied force on it?

Chapter 10 Gravitation

Section 10-1 Gravitation is a force of universal importance

Goal: Identify what it means to say that gravitation is universal.

Concept Check

1. Two objects are moved apart at a constant rate so that the distance between them increases. Does the gravitational force acting between them increase, decrease, or stay the same? Explain your answer.

2. The Moon is held in its orbit by the gravitational force between the Moon and Earth. Is there a gravitational force between your body and the Moon? Explain your answer.

Gravitation

Section 10-2 Newton's law of universal gravitation explains the orbit of the Moon

Goal: Explain how Newton's law of universal gravitation describes the attractive gravitational force between any two objects.

Concept Check

1. What are the two physical quantities that determine the magnitude of the gravitational force two objects exert on each other?

2. Describe the condition under which the formula $w = mg$ is a reasonable approximation.

Problem-Solving Review

We will find the degree to which the approximation $w = mg_{moon}$ is valid. An object with mass 13.0 kg is resting on the surface of the Moon. The Moon has a mass of 7.35×10^{22} kg and a radius of 1.74×10^6 m.

Set Up

1. Use Newton's law of gravitation to find the force of the Moon on the object.

2. What is the acceleration due to gravity on the surface of the Moon, g_{moon}?

Solve

1. How far from the surface of the Moon would the object have to be raised so that the gravitational force would be 1.00% less than what it was initially?

2. Up to what distance from the Moon will the magnitude of the force of gravity using Newton's universal law of gravitation differ from the result obtained from $w = mg_{moon}$ by less than or equal to 39.0%?

Reflect

1. If the mass of the Moon were doubled, how would this change the force exerted on the object?

2. What is the gravitational force of the object on the Moon?

Section 10-3 The gravitational potential energy of two objects is negative and increases toward zero as the objects are moved farther apart

Goal: Describe the general expression for gravitational potential energy and how to relate it to the expression used near Earth's surface.

Concept Check

1. An alien holds two bowling balls. How far apart does he have to hold them in order for each one not to be affected by the force of gravity due to the presence of the other?

2. Would the escape velocity be greater than, less than, or the same on the Sun compared to Earth? Explain.

Problem-Solving Review

A 30.0 kg meteorite is approaching Earth with a speed of 37.0 km/s at a distance of 3 times the radius of Earth. It loses 1.30 gigajoules due to friction with the atmosphere. Let's calculate the final speed of the meteorite just before it impacts Earth's surface. (The radius of Earth is approximately 6300 km, and its mass is 5.97×10^{24} kg.)

Set Up

1. What is the meteorite's distance to Earth in km when it is 3 Earth radii away?

2. Write an expression for the total initial energy of the meteorite in terms of the meteorite's initial potential energy U_0 and initial kinetic energy K_0. Write another expression for the total final energy of the meteorite in terms of the final potential energy U_f and final kinetic energy K_f.

3. Using the expressions you found in the previous question, write an expression of the conservation of energy in terms of U_0, K_0, U_f, K_f, and energy lost to friction, W_f.

Solve

1. Expand the expression you found in the previous question by writing out the potential and kinetic energies in terms of the meteorite's mass and speed and Earth's mass and radius.

2. Solve the equation you found in the previous question to find an expression for the final speed of the meteorite.

3. Use the given quantities to find the final speed of the meteorite.

Reflect

1. What is the final kinetic energy of the meteorite?

2. A distance d lies between the meteorite and Earth. Earth exerts a gravitational force of F_{Grav} on the meteorite resulting in a gravitational potential energy U_{Grav}. The meteorite moves toward Earth such that now the distance between them is $d/3$. What is the force of gravity between them now, in terms of F_{Grav}? What is the potential energy due to this force, in terms of U_{Grav}?

Section 10-4 Newton's law of universal gravitation explains Kepler's laws for the orbits of planets and satellites

Goal: Apply the law of universal gravitation and the expression for gravitational potential energy to analyze the orbits of satellites and planets.

Concept Check

1. Kepler's third law states that the square of the orbital period is proportional to the cube of the semimajor axis. Describe the difference between two quantities that are proportional and two quantities that are equal.

2. What is the relation between the orbital speed of an object and its orbital radius? What about the orbital speed and the orbital period?

Problem-Solving Review

Saturn's orbit has a semimajor axis of 1.43×10^{12} m. The Sun has a mass of 1.99×10^{30} kg, and the mass of Saturn is 5.68×10^{26} kg. A NASA satellite with a mass of 200 kg is sent in circular orbit 15.0 km above Saturn to study its properties. (Saturn has a radius of about 60,000 km.) When Saturn is closest to the Sun, it is 1.35×10^{9} km away.

Set Up

1. Sketch Saturn's orbit around the Sun. Label the semimajor axis.

2. Write an expression for the velocity of a satellite in circular orbit in terms of Saturn's mass and orbital radius.

Solve

1. What is the gravitational potential energy of Saturn when it is at the semimajor axis and closest to the Sun?

2. What is the velocity of the NASA satellite? What is its kinetic energy?

3. What is the gravitational potential energy of the NASA satellite?

4. What is the total mechanical energy of the NASA satellite?

5. What is the orbital period of the NASA satellite about Saturn?

Reflect

1. Using the fact that Saturn is 9.02 AU (astronomical units) from the Sun and Earth is 1.00 AU from the Sun, find the orbital period of Saturn.

2. Which of Kepler's three laws can be explained by Newton's law of universal gravitation?

Section 10-5 Apparent weightlessness can have major physiological effects on space travelers

Goal: Explain the origin of apparent weightlessness.

Concept Check

1. A space shuttle orbits Earth at a distance of 400 km, where the crew feels the effects of apparent weightlessness. If they reduce their height to 350 km, will their acceleration due to gravity increase, decrease, or stay the same?

Problem-Solving Review

A space station orbits Earth at a height of 375 km. The entire station has a mass of 5.00×10^5 kg. Alice, a crew member with a mass of 55.0 kg, stands inside the space station. Let's explore the effects of weightlessness inside the space station.

Set Up

1. On the sketch below indicate with solid lines and arrows the forces acting on Alice. Indicate with dotted lines the forces acting on the space station.

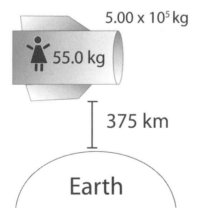

2. Write an expression for the force of gravity on Alice.

3. Write an expression for the force of gravity on the space station.

Solve

1. Use Newton's second law to find an expression for the acceleration of Alice due to gravity in terms of her mass and the force of gravity you found in question 2 of the Set Up.

2. Use Newton's second law to find an expression for the acceleration of the space station due to gravity in terms of its mass and the force of gravity you found in question 3 of the Set Up.

3. Explain why Alice feels weightless.

Reflect

1. Assuming the space station follows a circular orbit, what is its linear velocity?

Chapter 11 Fluids

Section 11-1 Liquids and gases are both examples of fluids

Goal: Describe the similarities and differences between liquids and gases.

Concept Check
1. What is the difference between a fluid and a solid?

2. What is the difference between a gas and a liquid?

3. Why are the attractive forces between molecules in a liquid stronger than in a gas?

Section 11-2 Density measures the amount of mass per unit volume

Goal: Recognize how to apply the definition of density.

Concept Check
1. Describe the difference between density and mass.

2. Explain the relationship between specific gravity and density.

Problem-Solving Review
You are studying an unknown metal. The piece of metal forms a cylinder of height 10.00 cm and radius 4.00 cm. To find the mass of the cylinder you hang it from a spring with a spring constant of 9517 N/m. When the cylinder is placed on the spring it stretches 0.404 cm. Let's find out what kind of metal the cylinder is.

Set Up
1. Draw a free-body diagram of the cylinder hanging from the spring.

2. Write an expression for the volume of the cylinder in terms of the given dimensions. What is the volume of the cylinder?

Solve
1. Write Newton's first law for the cylinder.

2. Solve the expression you found in the previous question for the mass of the cylinder.

3. What is the density of the cylinder?

4. Use equation 11-1 in the textbook to determine what kind of material the cylinder is made of.

Reflect
1. If the cylinder was compressed down to half its volume, what would its density be?

Section 11-3 Pressure in a fluid is caused by the impact of molecules

Goal: Explain the origin of fluid pressure in terms of molecular motion.

Concept Check
1. What is pressure? How is pressure different than force?

2. What is the equation that defines pressure? Based on this equation, what are the units of pressure?

Problem-Solving Review
It is possible to crush a steel 55.0 gallon storage drum using air pressure by evacuating some of the air inside. Gallons are units of volume, and one gallon is 4000 cm^3. Fifty-five gallon drums typically have a radius of 0.310 m. Atmospheric pressure is about 10^5 N/m^2.

Set Up
1. The drum is the shape of a cylinder. What is the area of the bottom of the drum in m^2?

2. What is the height of the drum?

3. What is the surface area of the drum? (*Hint*: Remember there are three regions to the drum—the top, the bottom, and the sides.)

Solve

1. Find the total force from atmospheric pressure on the drum.

2. Now if you evacuate the air from inside the drum so the total pressure inside the drum is 5.00×10^4 N/m³, there will be a net force exerted from the air on the outside. What is the net force on the drum from the air remaining inside it?

3. If the force required to bend steel of this thickness is on the order of 10^4 N, what will happen to the drum?

Reflect

1. What quantity can be changed to decrease the force acting on an object from pressure?

2. If you placed an object inside the drum before it collapsed, what would be the pressure on that object?

Section 11-4 In a fluid at rest, pressure increases with increasing depth

Goal: Calculate the pressure at a given depth in a fluid in hydrostatic equilibrium.

Concept Check

1. Explain why pressure increases with depth in a still-standing fluid.

2. If you drink from a water bottle on an airplane and then keep the empty water bottle sealed while the plane lands, the bottle will collapse. Explain why this happens.

Problem-Solving Review

A steel submarine can support pressure of up to 4.00×10^6 Pa on its outer hull. Under this condition, let's see to what depth one of these submarines can submerge. Assume that atmospheric pressure is 1.01×10^5 Pa and that the density of water is 1000 kg/m^3.

Set Up

1. As the submarine dives, does the pressure on its hull increase, decrease, or stay the same? Why?

2. What is the pressure exerted on the submarine while it is surfaced?

Solve

1. Write an expression for the pressure on the submarine while it is submerged to a depth d, in terms of the pressure on it while it is surfaced p_0, the density of water and the gravitational constant.

2. Solve the equation you found in the previous question for the depth d.

3. What is the maximum depth to which the submarine can be submerged?

Reflect

1. How far below the surface would the submarine have to dive to exactly triple the pressure on its hull?

Chapter 11

Section 11-5 Scientists and medical professionals use various units for measuring fluid pressure

Goal: Explain the difference between absolute pressure and gauge pressure.

Concept Check

1. What is the difference between absolute pressure and gauge pressure?

2. Is blood pressure an absolute pressure or a gauge pressure? Explain.

Problem-Solving Review

At a visit to the doctor, you get your blood pressure taken and the results are 95/75 mm/Hg, where the top number is the systolic pressure and the bottom number is the diastolic. The density of mercury is 1.36×10^3 kg/m^3.

Set Up

1. What is atmospheric pressure in mmHg?

Solve

1. What is the systolic pressure in N/m^2?

2. What is the diastolic pressure in N/m^2?

3. Blood pressure is a gauge pressure. What are the values of the absolute systolic and diastolic pressure in Pascals?

Reflect

1. If your blood pressure were taken on a planet where the atmospheric pressure was the same, but the value of g was twice what it was on Earth, what would the value be in mmHg?

Chapter 11

Section 11-6 A difference in pressure on opposite sides of an object produces a net force on the object

Goal: Calculate the force on an object due to a difference in pressure on its sides.

Concept Check

1. Explain why it is the difference in pressures that results in a net force.

2. The engine in Car A has pistons that have twice the surface area of those in Car B. If the pistons in each car are exerting the same amount of force, which car's pistons are under a greater amount of pressure? Explain.

Problem-Solving Review

Lisa inflates a birthday balloon such that the pressure inside is 16.0 psi (lb/in²) and lets it float off up into the sky. Let's calculate how high the balloon can travel before the pressure on its surface causes it to pop. The balloon is shaped like a sphere with radius 72.0 cm and can withstand a net force of 5.02×10^4 N before popping. (Assume the size of the balloon doesn't change.)

Set Up

1. What is the surface area of the balloon in m²?

2. Write an expression for the force acting on the balloon's surface in terms of the pressure difference between the inside and outside of the balloon and the balloon's surface area. Will the direction of the force be pointing toward the inside of the balloon or pointing out of the balloon?

3. What is the initial internal pressure of the balloon in Pascals?

Solve

1. What is the pressure on the inside and outside of the balloon before it is let go?

2. What is the pressure on the inside of the balloon when it has reached its maximum height?

3. What is the pressure on the outside of the balloon the instant before it pops?

4. Write an expression for the atmospheric pressure on the balloon when it has reached its maximum height in terms of the atmospheric pressure on the ground, the density of air (1.23 kg/m³), Newton's constant of gravitational acceleration g, and the maximum height.

4. Solve the expression you found in the previous question for the maximum height the balloon can reach.

Reflect

1. Once you know the pressure, what (if any) effect is there from the composition of the gas inside the balloon on the force inside the balloon?

Section 11-7 A pressure increase at one point in a fluid causes a pressure increase throughout the fluid

Goal: Explain how to apply Pascal's principle to a fluid at rest.

Concept Check
1. What is Pascal's principle?

2. Briefly explain how a hydraulic jack works.

Problem-Solving Review
On a trip to the mechanic, your 2000 kg car is lifted 2.00 m with a hydraulic jack. The size of the large piston with the hydraulic fluid lifting the car has a radius of 0.500 m. The small piston has a radius of 0.300 m.

Set Up
1. What is the minimum force that needs to be placed on the bottom of the car to lift it?

2. What is the area of each piston?

Solve
1. What is the minimum force on the small piston required to lift the car?

2. What is the gauge pressure on the head of each of the cylinders?

3. How far do you have to push the small cylinder down to raise the car 2.00 m?

Reflect

1. What is the work done on the small piston to lift the car 2.00 m?

2. What is the work done by the large piston to lift the car 2.00 m?

Section 11-8 Archimedes' principle helps us understand buoyancy

Goal: Apply Archimedes' principle to find the buoyant force on an object in a fluid.

Concept Check

1. Explain how a submarine with a steel hull of much greater density than water is able to float.

2. Explain the difference between apparent weight and true weight.

Problem-Solving Review

You need to send a piece of steel down a river. You decide to float it down on a block of wood you have. The block of wood is 120.0 cm long, 30.0 cm wide, and 20.0 cm high and has a density of 300 kg/m³. Let's find out the maximum mass of steel that can be placed on top of the wooden block before it becomes halfway submerged. The density of water is 1000 kg/m³.

Set Up

1. Draw a free-body diagram of all the forces acting on the wooden block.

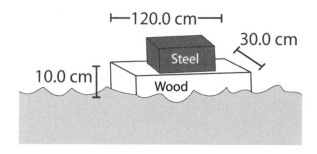

2. Write an expression for the buoyant force on the wooden block, assuming it is halfway submerged.

3. Write an expression for the force of gravity on the wooden block in terms of its density and volume.

Solve

1. Write Newton's first law for the wooden block.

2. Solve the expression you found in the previous question for the mass of steel m_s that can be placed on the block, if the wooden block is halfway submerged.

Reflect

1. What is the maximum mass of steel that can be placed on the wooden block before the entire wooden block is submerged?

2. If the density of the steel is 8000 kg/m³, what is the maximum volume of steel that can be placed on top of the wooden block?

Section 11-9 Fluids in motion behave differently depending on the flow speed and the fluid viscosity

Goal: Use the equation of continuity to analyze the flow of an incompressible fluid.

Concept Check
1. What is viscosity?

2. Describe the difference between turbulent and laminar flow. Give an example of each.

3. Describe the difference between viscous and inviscid flow. Give an example of each.

Problem-Solving Review
In the near future you land a great job as an engineer for a drinking fountain company. You are tasked with designing a part for a drinking fountain pump that can pump out water with a maximum velocity of 2.00 m/s. Unfortunately, you don't know how big to make the adapter (r_1). The water is supposed to exit the fountain from a hole 1.00 cm across at an angle of 45.0° and reach a maximum height of 7.50 cm. (Typically frictional forces would prohibit this type of analysis from working, but we will neglect them for this exercise.)

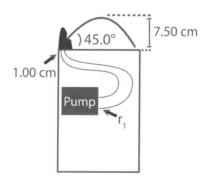

Set Up

1. What is the velocity that the water needs to leave the drinking fountain with (this can be approximated as a projectile motion problem)? Call this quantity v_2.

2. What is A_2?

3. What is the volume flow rate for the water leaving the fountain?

Solve

1. Assuming the pump should generally be operating at 75.0% capacity, what will the velocity of the water leaving the pump be?

2. Find r_1, the size of the radius of the adapter needed for the fountain to work correctly.

Reflect

1. Describe a problem with this technique you could encounter if you were designing a system that was pumping air instead of water.

Section 11-10 Bernoulli's equation helps us relate pressure and speed in fluid motion

Goal: Apply Bernoulli's principle to relate fluid pressure and flow speed in an incompressible fluid.

Concept Check

1. Which properties must a fluid have in order to satisfy Bernoulli's equation?

2. Describe the relationship between pressure and velocity in a moving fluid.

3. Explain what makes a curveball curve.

Problem-Solving Review

Water is pumped into a 10-story apartment building. The water comes into the apartment building at ground level through a pipe of radius 3.00 cm at a velocity of 2.00 m/s. We will calculate the pressure of the water entering the apartment building, reaching the 10th floor at a height of 30.0 m, and flowing out of a water tap of radius 2.00 cm. (Assume atmospheric pressure of 1.00 atm and the density of water is 1000 kg/m³.)

Set Up

1. What is the cross-sectional area of the pipe through which the water enters the building? What is the cross-sectional area of the water tap?

2. Use the equation of continuity to find the velocity at which the water will exit the tap at the 10th floor.

3. What is the minimum pressure of the water exiting the tap?

Solve

1. Write Bernoulli's equation for the water at the bottom and top of the apartment building.

2. Solve Bernoulli's equation for the pressure needed at the bottom of the apartment building.

3. What is the pressure needed at the bottom of the apartment building in atm?

Reflect

1. What is the gauge pressure at the bottom of the apartment building?

Section 11-11 Viscosity is important in many types of fluid flow

Goal: Explain what happens in flows where viscosity is important.

Concept Check
1. Describe what the Reynolds number is in words.

2. What are characteristics of a fluid with a high Reynolds number? What about a low Reynolds number?

Problem-Solving Review
A syringe is 15.00 cm long and has a radius of 1.00 cm. The needle at the end has a radius of 0.080 cm and a negligible length. As an experiment (assume all safety precautions are taken!) you fill the syringe with mercury and exert a force of 5.00 N on the plunger. The plunger moves 1.00 cm/s and pushes some air out of the syringe. For clearing the air out of the syringe we can consider the system as a tube with laminar flow. The viscosity of mercury is 1.55×10^{-4} Ns/m².

Set Up
1. What is the volumetric flow rate for the syringe?

2. What is the gauge pressure on the plunger?

Solve

1. What is Reynolds number for the mercury in the syringe?

2. What is the change in pressure from the plunger to the needle end of the syringe?

Reflect

1. How would switching to a fluid that had twice the viscosity of mercury change the pressure at the end of the syringe?

Section 11-12 Surface tension explains the shape of rain drops and how respiration is possible

Goal: Describe the role of surface tension in the behavior of liquids.

Concept Check

1. Explain why molecules in a liquid arrange themselves to minimize the liquid's surface area?

2. Explain how the surface tension in a water droplet changes with the size of the droplet?

Chapter 12 Oscillations

Section 12-1 We live in a world of oscillations

Goal: Define oscillation.

Concept Check

1. Describe the relationship between an object's oscillatory motion and equilibrium position. How is motion around an equilibrium position different than the kind of motion we studied in previous chapters?

2. Explain how a force can cause oscillatory motion.

Section 12-2 Oscillations are caused by the interplay between a restoring force and inertia

Goal: Describe the key properties of oscillations, including what makes oscillation happen.

Concept Check

1. Describe in words what is meant by the amplitude of an oscillation.

2. What is the difference between period and frequency? What are the units of period and frequency?

3. What is the period of Earth about the Sun? What is the period of Earth about its axis?

Problem-Solving Review

While taking a sea voyage you set a glass of coffee on your desk. Due to the motion of the ship, each time the ship bobs in the water an average of 0.600 ml of coffee spills out of your cup. (The coffee spills out when the ship reaches its maximum height of bobbing and starts to fall back down.) It takes 43.0 minutes for the cup to empty.

Set Up

1. A standard cup of coffee is about 240 ml. How many spills have to occur for the cup to be empty?

Solve

1. Assuming they occur at a steady rate, what is the amount of time between each spill?

2. What is the period of the ship's bobbing?

3. What is the frequency of the ship's bobbing?

4. If the total distance the ship travels from the bottom of its oscillation to the top is 10.0 m, what is the amplitude of the ship's motion?

Reflect

1. Describe the forces that are causing the ship oscillation?

2. What provides the restoring force for the ship's motion?

Section 12-3 The simplest form of oscillation occurs when the restoring force obeys Hooke's law

Goal: Explain the connection between Hooke's law and the special kind of oscillation called simple harmonic motion.

Concept Check

1. A mass attached to a spring oscillates back and forth, obeying Hooke's law. Explain why the frequency at which the mass oscillates is independent of how far the spring was initially stretched.

2. You pull back a mass on a spring, let it go, and it starts to oscillate. Explain at which positions the velocity of the mass is the greatest and the least.

Problem-Solving Review

A mass m is attached to a spring with spring constant of 200 N/m. It is oscillating with an angular frequency of 200 rad/s and phase angle $\pi/4$. The maximum velocity the mass reaches is 0.500 m/s. Let's explore its motion.

Set Up

1. Write an expression for the mass m in terms of the spring constant k and the angular frequency ω. How large is the mass?

2. Write down expressions for the position, velocity, and acceleration of the mass as sinusoidal functions in terms of time t, maximum amplitude A, angular frequency ω, and phase angle ϕ.

$x =$

$v =$

$a =$

Solve

1. Write an expression for the amplitude A of the motion of the mass in terms of its maximum velocity v_{max} and its angular frequency ω. (*Hint*: The maximum and minimum values of $\sin(\omega t + \phi)$ are 1 and −1). What is the amplitude A?

2. Sketch the position of the mass as a function of time.

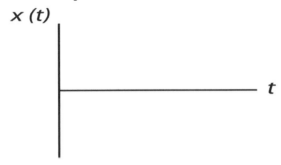

3. Sketch the velocity of the mass as a function of time.

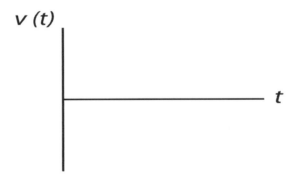

4. Sketch the acceleration of the mass as a function of time.

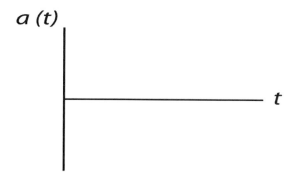

Reflect
1. How would the maximum velocity of the mass change if the phase angle were doubled?

2. How would the sketches you drew in parts 2–4 of the Solve section look different if the phase angle were shifted by 2π?

Section 12-4 Mechanical energy is conserved in simple harmonic motion

Goal: Discuss how kinetic energy and potential energy vary during an oscillation.

Concept Check

1. A mass on a spring laying horizontally on a frictionless surface is pulled back and then released to oscillate freely. Where does it have the most potential energy?

2. For the case above, where is the kinetic energy of the block the highest?

3. Describe in words the evolution of the total energy of the system.

Problem-Solving Review

During a stint as a biologist you wish to measure the spring constant of a spider's silk. To do this, you have a spider attach its string to a surface and then you pull the spider some distance from equilibrium on a slippery horizontal surface. You pull the spider with a mass of 20.0 g a distance of 13.0 cm from equilibrium and let go. The spider arrives back at the equilibrium position 0.400 s later.

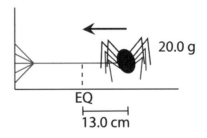

Set Up
1. Where will the spider have its maximum velocity?

2. Calculate the maximum velocity of the spider.

Solve
1. Find the maximum kinetic energy of the system. At what x position does this occur?

2. Use the maximum kinetic energy to find the total energy of the system. What is the maximum potential energy of the system?

3. What is the spring constant for the spider silk?

Reflect
1. Describe any changes you would encounter if you conducted this experiment with the spider hanging vertically.

Section 12-5 The motion of a pendulum is approximately simple harmonic

Goal: Explain what determines the period, frequency, and angular frequency of a simple pendulum.

Concept Check
1. Explain why we are able to use the same equations for the motion of a pendulum as we did for the motion of a mass attached to a spring.

2. Explain why for small angles the frequency of a pendulum does not depend on its initial displacement from equilibrium.

Problem-Solving Review
John makes a pendulum by tying a rope from his ceiling to a 7.00 kg bowling ball. John's mass is 100 kg. He climbs on to the bowling ball and swings back and forth with a period of 3.50 seconds. Let's explore his motion. (Assume for this problem both the mass of the bowling ball and the mass of John are concentrated at the center of the bowling ball.)

Set Up
1. Write an expression for the angular frequency of the pendulum in terms of its period. What is the pendulum's angular frequency?

2. Write an expression for the torque about the point where the pendulum is attached to the ceiling in terms of the mass on the pendulum m, the length of the pendulum L, the angle θ, and the gravitational constant g.

Solve
1. Write an expression for the length of the pendulum L in terms of its angular frequency ω and the gravitational constant g. What is the length of the pendulum?

2. Find the torque around the point where the rope is connected to the ceiling, when the angle of the rope is 5.00°.

3. If the length of the rope were to be doubled, what would be the frequency f of the pendulum?

Reflect
1. How will the angular frequency change if John climbs off the bowling ball and lets the ball swing by itself? How will the torque change? (Assume the location of the center of mass remains at the center of the bowling ball.)

Section 12-6 A physical pendulum has its mass distributed over its volume

Goal: Explain what determines the period, frequency, and angular frequency of a physical pendulum.

Concept Check

1. What determines the period, frequency, and angular frequency of a physical pendulum?

2. Explain why the physical pendulum cannot be modeled as a point mass.

Problem-Solving Review

A poster of mass 500 g hangs by a tack on the wall. The poster is 40.0 cm across, 65.0 cm high, and is held up by a nail in the center of the poster. The nail is positioned close to the top of the poster. A gust of wind disturbs the poster, causing it to swing before returning to its equilibrium position after 0.800 s.

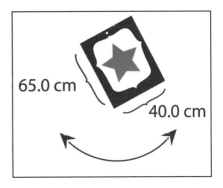

Set Up

1. Explain why you cannot consider the mass of the poster as concentrated in a single point.

2. What is the period of oscillation of the poster? Its frequency?

3. What is the distance of the pivot to the poster's center of mass?

Solve
1. Find the angular velocity ω for the poster.

2. What is the moment of inertia for the poster?

Reflect
1. What is the period for a pendulum with mass 0.500 kg and a center of mass the same distance from the nail as the posters?

2. How does this compare to the period of the poster?

Section 12-7 When damping is present, the amplitude of an oscillating system decreases over time

Goal: Describe what happens in underdamped, critically damped, and overdamped oscillations.

Concept Check
1. Explain the difference between underdamped, critically damped, and overdamped oscillations.

2. Three similar systems—one underdamped, one overdamped, and one critically damped—start oscillating at the same time. Describe their motion after an amount of time much greater than their periods.

Problem-Solving Review
A mass of 3.00 kg is attached to a spring with spring constant 300 N/m and submerged in a viscous liquid that causes the spring's oscillations to be underdamped. After being released, the amplitude of the spring's motion decreases by 30.0% in two seconds. Let's calculate the damping constant b and discover some other properties of the spring's motion.

Set Up
1. Write an expression for the amplitude $A(t)$ as a function of time.

2. Rearrange the expression you found in the previous question, solving for the damping constant b.

3. What is the ratio of the amplitude after two seconds $A(t = 2\text{s})$ to the amplitude A at time $t = 0$? (Give a numerical value.)

Solve

1. Using the expression you found in question 2 of the Set Up and the ratio you found in question 3, find the numerical value of the damping constant b.

2. Write an expression for the damped angular frequency of the spring ω_{damped}. What is the damped angular frequency of the spring ω_{damped}?

3. What is the damped period T?

4. Now imagine that we change the fluid in which the spring is sitting, in order to increase b until the spring is critically damped. What is the damped angular frequency of the spring ω_{damped} if the spring is now critically damped?

5. Find an expression for the new damping constant b in terms of the spring constant k and the mass m. What is the new damping constant b?

Reflect

1. For the first case when the spring was underdamped, how many times does the spring pass the equilibrium position before its amplitude is decreased by 30.0%?

Section 12-8 Forcing a system to oscillate at the right frequency can cause resonance

Goal: Identify the circumstances under which resonance occurs in a forced oscillation.

Concept Check

1. What are the conditions where resonance occurs in a forced oscillation?

2. When you play a low C on a piano, you can hear a sound from the high C even though you didn't press that key. Explain this phenomenon.

Problem-Solving Review

Consider the horizontal mass spring system with a driving force of 50.0 N. The mass is 3.00 kg and attached to a spring with a spring constant $k=75.0$ kg/s^{-2}. Assume friction acts in a way that the system is critically damped.

Set Up

1. Find the period and the natural frequency ω_0 of the mass.

2. Find the un-driven amplitude A.

3. Write the relation between k, m, and b for a critically damped system.

Solve

1. Write the equation for the amplitude of a driven oscillator with critical damping, in terms of k, m, ω_o, ω, and F_o.

2. Find ω.

3. What is A for the case where $\omega = \omega_o$?

4. What should ω be so that the amplitude of the driven oscillation is 40.0% greater than the damped, undriven case? (*Hint*: Make a ratio between the driven and undriven case, then solve for ω.)

Reflect

1. What would eventually happen if the system were underdamped?

2. What would be the force required to make the system oscillate with an $\omega = 30 \frac{rad}{s}$?

Chapter 13 Waves

Section 13-1 Waves are disturbances that travel from place to place

Goal: Describe what a mechanical wave is.

Concept Check
1. What is a mechanical wave?

2. You drop an object in the water at the beach and waves bob the object up and down. The motion of the object approximates the general motion of the water molecules nearby it. Explain why the object doesn't move to shore with the waves.

Section 13-2 Mechanical waves can be transverse, longitudinal, or a combination of these

Goal: Explain the key properties of transverse, longitudinal, and surface waves.

Concept Check

1. Explain the difference between transverse and longitudinal waves.

2. Explain what causes a wave to propagate.

3. For the following waves check the box for whether the wave is longitudinal or transverse (it might be both!):

	Longitudinal	Transverse
Ocean waves		
Sound waves		
Waves along a rope being shaken		
Earthquakes		

Section 13-3 Sinusoidal waves are related to simple harmonic motion

Goal: Define the relationship between simple harmonic motion and what happens in a sinusoidal wave.

Concept Check
1. What type of function typically describes SHM?

2. Explain what happens to the speed and wavelength of a wave when the frequency changes but the medium stays the same.

Problem-Solving Review
Consider the plot below of an object experiencing SHM with a velocity of 3.00 m/s:

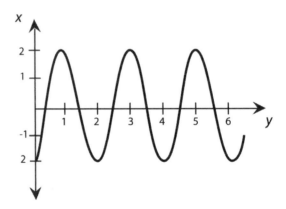

Set Up
1. Label the wavelength of the cosine wave on the plot above. What are the wavelength and wavenumber?

2. Approximately what are the period and frequency of the sinusoidal wave?

3. Recall that the graph shows a cosine function. What is an acceptable value of ϕ?

4. What is the maximum value of A?

Solve

1. Write $y(x,t)$ using the values you calculated for ω, k, ϕ, and A.

2. Find $y(1.2,1.0)$.

3. Sketch $y(x,\frac{\pi}{2})$ vs. x for $x=1,2,3,4$ at $t=\frac{\pi}{2}$.

4. Sketch $y(x, \pi)$ vs. x for $x=1,2,3,4$

Reflect

1. For the two sketches above, how do the graphs change with time?

2. What effect would increasing k have on the two plots?

Section 13-4 The propagation speed of a wave depends on the properties of the wave medium

Goal: Explain what determines the propagation speed of a mechanical wave.

Concept Check

1. Explain what determines the propagation speed of mechanical waves.

2. Will sound travel faster in a block of steel or a block of wood? Explain.

Problem-Solving Review

Zoe works in a forensics lab where she has to determine the chemical composition of several unknown samples. She decides to measure the speed of sound in each sample to try and determine the sample's properties.

Set Up

1. Write down two expressions, one for the speed of sound in a fluid and the other for the speed of sound in a solid.

Solve

1. The first sample is a liquid. Zoe measures the density of the liquid to be 0.827 g/cm^3 and the speed of sound in the liquid to be 1.10×10^3 m/s. What is bulk modulus of this fluid?

2. What is the identity of this fluid? (*Hint*: Check Table 9-1.)

3. The second sample is a solid. Zoe measures the density of the solid to be 9.00 g/cm³ and speed of sound through the solid to be 3333 m/s. What is the Young's modulus of this solid?

4. What is the identity of this solid?

Reflect

1. If the solid in question 3 of the Set Up were to be heated to a very high temperature, how would the speed of sound in the solid change?

Section 13-5 When two waves are present simultaneously, the total disturbance is the sum of the individual waves

Goal: Describe what happens when two sinusoidal waves from different sources interfere with each other.

Concept Check

1. Sketch a wave with the same frequency as the one below, but with a phase shift such that the superposition of the two waves makes the maximum amplitude 0.

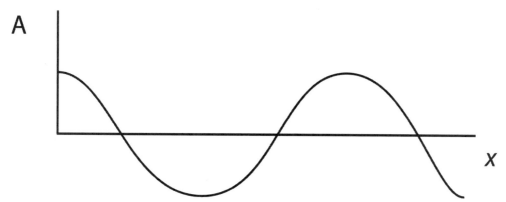

2. Sketch a wave below such that the two waves have the same frequency, but the maximum amplitude of the superposition of waves is A/2.

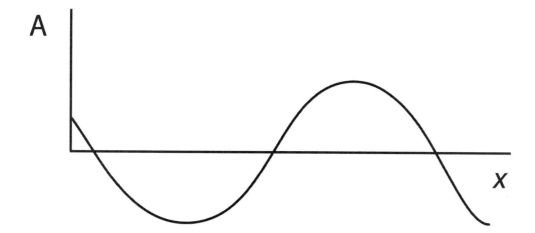

Problem-Solving Review

You are placing seats in a theater and want to determine which frequencies will interfere when played by two speakers 10.0 m apart. The first seat is placed 15.0 m back and 2.50 m from the wall. (Assume the speed of sound is 343 m/s.)

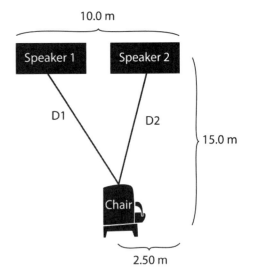

Set Up
1. Find the distance D_1 to speaker 1.

2. Find the distance D_2 to speaker 2.

3. Find the path length difference between D_1 and D_2.

Solve
1. Find three frequencies for which the sound waves from the speakers completely constructively interfere at the chair.

2. Find three frequencies for which the sound waves from the speakers completely destructively interfere at the chair.

Reflect

1. Imagine that you are listening to music through two stereo speakers. Explain why you do not usually find spots in a room where there is no sound.

Section 13-6 A standing wave is caused by interference between waves traveling in opposite directions

Goal: Define the properties of standing waves on a string that is fixed at both ends.

Concept Check
1. Explain how standing waves arise.

2. Describe which parameters determine the frequency of a standing wave on a string.

Problem-Solving Review
There is a 1.00 m long steel clothes line stretched taut between two apartment buildings. The line has a linear mass density of 2.00×10^{-3} kg/m, and the tension in the line is 117 N. Let's explore the properties of the standing waves on this line.

Set Up
1. Write an expression for the possible frequencies of standing waves on the clothes line in terms of the length L, the tension F, the linear mass density μ, and the mode integer n.

2. Write an expression for the possible wavelengths of standing waves on the clothes line in terms of the length L and the mode integer n.

Solve

1. What is the frequency of the first harmonic?

2. What is the wavelength of the wave on the string for the first harmonic?

3. What is the wavelength of the sound wave in the air produced by the first harmonic of the clothes line? (Assume the speed of sound in air is 343 m/s.)

4. Sketch the line oscillating in the third harmonic ($n = 3$).

5. What are the positions of the nodes in the third harmonic? What are the positions of the antinodes?

Reflect

1. If the length of the clothes line were doubled, what would happen to the wavelength of the sound wave in the air produced by the first harmonic?

2. What would the wavelength be if the tension were doubled?

Section 13-7 Wind instruments, the human voice, and the human ear use standing sound waves

Goal: Explain the nature of standing sound waves in open and closed pipes.

Concept Check

1. Give an example of a musical instrument closed at one end and open at the other, and of a musical instrument open at both ends.

2. What effect does lengthening the pipe have on the frequency?

3. If a standing wave were set up in a closed pipe (a pipe with one open end and one closed end) and then the closed side were suddenly removed, would the frequency of the sound from the pipe go up or down?

Problem-Solving Review

You are designing an organ for an organ-making company. Typically organs can be played with open or closed pipes. You wish to calibrate the fundamental frequency to C, which is often considered to be 256 Hz. Let's calculate the effect of different organ designs on the sound of the organ. (Assume the speed of sound in air is 343 m/s.)

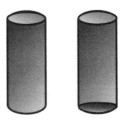

Set Up
1. What is the wavelength of middle C?

Solve
1. What is the minimum pipe length required to play middle C in an open pipe?

2. What is the minimum length required to play middle C in a closed pipe (one end is closed and one end is open)?

3. Assume the organ has open and closed pipes of the lengths you found in the previous questions. What would the frequency be of the second available harmonic in the open pipe? In the closed pipe?

4. Sketch the second available harmonic for both the open and closed pipes. Label nodes and anti-nodes.

Reflect
1. For the open and closed pipe above, what are two frequencies that are inaccessible to the open pipe but can be played on the closed pipe?

Section 13-8 Two sound waves of slightly different frequencies produce beats

Goal: Describe how beats arise from combining two sound waves of slightly different frequencies.

Concept Check
1. Explain the conditions under which the beat phenomenon arises.

2. Explain why you do not experience the beat phenomenon every time you hear two frequencies at the same time.

Problem-Solving Review
Tammy is tuning her guitar. The first string is correctly tuned to 329.6 Hz, but she hears it beat once every 3.00 seconds with the sixth string. Let's calculate how much tension she should add to the sixth string in order to correctly tune it to 82.4 Hz.

Set Up
1. Write an expression for the beat frequency in terms of the two interfering frequencies of the guitar.

2. Write an expression for a standing wave on a guitar string in terms of the string's length L, the tension F, the linear mass density μ, and the mode number n.

Solve
1. Which harmonic of the sixth string is beating with the first string?

2. How large is the beat frequency?

3. Tammy tries to increase the tension on the sixth string, but that makes the time in between beats longer. At what frequency f_m is the mistuned string oscillating?

4. Calculate the ratio between the desired frequency of the sixth string and the mistuned frequency f_m. By what factor does Tammy need to change the tension in the string?

Reflect
1. If the linear mass density of the sixth string were smaller, would the beating interval be longer or shorter?

2. If the sixth string was a little bit shorter, would the beating interval be longer or shorter?

Section 13-9 The intensity of a wave equals the power that it delivers per square meter

Goal: Explain what is meant by the intensity and sound intensity level of a sound wave.

Concept Check

1. Explain the difference between the intensity and the power of a sound wave.

2. Describe the relationship between sound intensity and the distance between the source and the observer.

Problem-Solving Review

You are designing a tsunami warning system that needs to be at least 70 dB up to 1000 m away and emits a frequency of 1500 Hz. (Assume the speed of sound in air is 343 m/s and the density of air is 1.23 kg/m³.)

Set Up

1. What is the angular frequency of the omitted oscillations?

Solve

1. What is the sound intensity at 1000 m away?

2. Find the power required to run the warning system.

3. Find p_{max}, the pressure amplitude of the wave?

4. Find the amplitude of the sound wave from the warning system.

Reflect

1. How would doubling the frequency change the power required?

2. What would the power have to be if you wanted the sound level to be at 70 dB 2000 m away?

Section 13-10 The frequency of a sound depends on the motion of the source and the listener

Goal: Recognize why the frequency of a sound changes if the source and listener are moving relative to each other.

Concept Check

1. Explain the causes of the Doppler effect.

2. Explain why planes moving faster than the speed of sound produce a sonic boom.

Problem-Solving Review

A jet takes off in New York to fly to London. Sam lives near the airport. When the jet is 400 m from his house, he hears the frequency of the jet's engine to be different than its true frequency by a factor of 2. Assume the speed of sound in air is 343 m/s. Let's explore the consequences of the Doppler effect for observers on the ground.

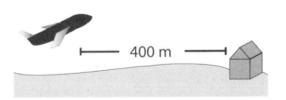

Set Up

1. As the jet approaches Sam's house, does he hear the jet's engine at its true frequency f, at a higher frequency, or at a lower frequency?

2. Write an expression for the frequency Sam observers f' in terms of the true frequency f of the jet engine, the speed of sound, and the speed of the jet.

3. What is the ratio of the jet's observed frequency f' to its true frequency f? (Give a numerical value.)

Solve

1. Rearrange the expression you found in question 2 to solve for the speed of the jet in terms of the ratio of the jet's observed frequency f' to its true frequency f and the speed of sound.

2. Use the expression you found in the previous question to calculate the speed of the jet at the time Sam observed it in units of km/hr.

3. When the Concorde reaches its full speed it travels faster than the speed of sound in air. Write an expression for the angle of the Mach cone formed by a supersonic jet.

4. Calculate the angle of the Concorde's Mach cone if the jet is traveling at 2100 km/hr.

Reflect

1. If an object travels the same speed in water as in air (a speed greater than the speed of sound both in air and in water), will its Mach cone angle be larger in air or water?

Chapter 14 Thermodynamics I

Section 14-1 A knowledge of thermodynamics is essential for understanding almost everything around you—including your own body

Goal: Define what thermodynamics is.

Concept Check

1. Explain what kinds of physical phenomena thermodynamics sets out to describe.

2. What are three important thermodynamic quantities?

Section 14-2 Temperature is a measure of the energy within a substance

Goal: Explain the meaning of temperature and thermal equilibrium.

Concept Check

1. What physical phenomenon does temperature measure?

2. Explain the condition for thermal equilibrium.

Problem-Solving Review

Let's explore different conditions of thermal equilibrium.

Set Up

1. Write the equation that converts temperatures in Fahrenheit to Celsius.

2. Write the equation that converts temperatures in Kelvin to Celsius.

Solve

1. Consider the following phenomenon and whether the two cases are in thermodynamic equilibrium or not.

Phenomenon	In thermodynamic equilibrium?(yes/no)
You and a room at room temperature (72°F)	
A stick and a piece of metal that have been untouched outside	
A glass of warm tea and the air in a refrigerator right after you put the tea in	
The same glass of warm tea that you put in the refrigerator and the air in the refrigerator a long time later	
Two gasses, gas A and gas B, where both have separately been allowed to come to thermal equilibrium with gas C	

Reflect

1. If you put a warm cup of coffee in a room, what happens to both of their temperatures?

2. Rank the following temperatures from highest to lowest:

0° Kelvin   0° Fahrenheit   0° Celsius

Section 14-3 In a gas, the relationship between temperature and molecular kinetic energy is a simple one

Goal: Describe the origin of pressure in an ideal gas, and explain the relationship between molecular kinetic energy and gas temperature.

Concept Check

1. Explain in terms of the motion of its constituent molecules why a gas in a container exerts pressure on the walls of the container.

2. Explain the meaning of degrees of freedom and the consequences of the equipartition theorem.

3. Describe which quantities affect the magnitude a molecule's mean free path.

Problem-Solving Review

A box with dimensions 20.0 cm × 50.0 cm × 30.0 cm is filled with 100 moles of nitrogen (which is diatomic in air) at 22.0°C. One mole of diatomic nitrogen has a mass of 0.028 kg. Let's explore some properties of this gas.

Set Up

1. What is 22.0° Celsius in Kelvin?

2. Write an expression for the pressure of the gas in terms of the volume it occupies, the ideal gas constant, the number of moles, and the temperature of the gas.

3. How many molecules of nitrogen are in the box?

4. What is the mass of one molecule of diatomic nitrogen?

Solve

1. What is the pressure of the gas?

2. Write an expression for and then calculate the root mean square speed of a molecule in the gas. (*Hint:* In air a molecule of nitrogen has two atoms.)

3. Write an expression for and then calculate the mean free path of a molecule in the gas. (Assume the radius of a nitrogen molecule is $\sim 2.00 \times 10^{-10}$ m.)

Reflect

1. If the length of the box were doubled, what would be the change in the mean free path?

2. If the temperature of the gas is now at 2.00°C, and the volume of the box remains the same, what would be the pressure of the gas?

Section 14-4 Most substances expand when the temperature increases

Goal: Explain how objects change in size when their temperature changes.

Concept Check

1. Does water expand or contract when its temperature changes near freezing? Explain.

2. Does water expand or contract when its temperature changes far above freezing? Explain.

Problem-Solving Review

The Khalifa Tower in the United Arab Emirates has a height of 829.8 m (that's nearly a half-mile tall!). The temperature difference between the sea level and the top is appreciable, and drops by about 6.50°C per 1000 m. A steel beam at the top has a length of 1.00 m. The coefficient of linear expansion α for steel is $1.30 \times 10^{-5} K^{-1}$.

Set Up

1. Assuming the temperature on the ground is 40.0 °C, what is the temperature at the top of the Khalifa Tower?

Solve

1. What is the difference in the length of the steel beam from the top of the tower to the bottom?

2. Now consider a 300 ml glass of water with $\beta=2.07\times10^{-4}\ K^{-1}$. What will be the change in volume of the water in the glass from the top of the building to the bottom?

3. A steel plate of length 0.250 m, width 0.170 m, and height 0.010 m has a hole cut in it with radius 0.005 m. If this steel plate is brought from the top of the tower to the bottom, what will be the change in the volume of the hole? (The ratio of the volume of the hole to the volume of the plate will be the same at the top as at the bottom.) Take $\beta = 3\alpha_{steel}$.

Reflect

1. Based on their coefficients, is water or steel more susceptible to a change in volume based on a temperature change?

Section 14-5 Heat is energy that flows due to a temperature difference

Goal: Examine the relationship between the quantity of heat that flows into or out of an object and the temperature change of that object.

Concept Check
1. Explain the difference between heat and temperature.

2. Describe the difference between internal energy and heat.

3. Explain why heat will flow between two objects of different temperatures that are placed in contact with each other.

Problem-Solving Review
Charlie is boiling 1.50 kg of water on an electric stovetop made up of copper coils. The stove provides a temperature of 350°F. As the water was heated, the power went out in Charlie's house when the water reached 170°F. Let's calculate the temperature of both the water and the coil after they have reached thermodynamic equilibrium. The coil has a mass of 0.500 kg. (Assume the system is in thermal isolation.)

Set Up
1. Find the temperatures of both the water and coil immediately after the power went out in Celsius.

2. Write an expression for the heat flowing out of the coil Q_{coil} in terms of the heat flowing into the water $-Q_{water}$.

3. What is the relation between final temperatures of the coil and the water?

Solve

1. Write an expression for the heat flowing out of the coil in terms of the change in its temperature, its specific heat, and its mass.

2. Write an expression for the heat flowing into the water in terms of the change in its temperature, its specific heat, and its mass.

3. Use the heat-flow relation you found in question 2 of the Set Up to combine the expressions in questions 1 and 2 of the Solve into a single equation involving the change of temperature of both objects, their specific heats, and their masses.

4. Solve the equation you found in the previous question for the final temperature of the water.

5. What is the final temperature of the water? Will the water boil? (The boiling point of water is 100°C. You can look up the specific heats of different materials in Table 14-3 in the textbook.)

Reflect

1. What would have been the initial temperature of the coil in order for the water to boil even after the power went out?

Section 14-6 Energy must enter or leave an object in order for it to change phase

Goal: Explain how heat must flow in order to cause a substance to change between the solid, liquid, and gas phases.

Concept Check
1. Why can't you make water 101°C at atmospheric pressure?

2. Explain the critical point and the triple point.

Problem-Solving Review
Christine decides to go on a 5.00 km run. While she is running, she burns 75 kcal per kilometer.

Set Up
1. How many joules of energy does Christine use on her run?

2. If 30.0% of the energy Christine expends is useful mechanical energy, how much thermal energy needs to be dissipated by sweating (the thermal energy lost by Christine when the sweat evaporates)?

Solve

1. In order to stay hydrated, you need to drink the same amount of water you lose by sweating. Based on the amount of energy that needs to be dissipated, how much water does Christine need to drink to stay hydrated? (See Table 14-4.)

2. Consider a 2.00 kg block of ice at −1.00°C. How much energy is required to bring it to 0.00°C and then melt the block? The information required to find this answer can be found in Tables 14-3 and 14-4.

3. For the same block, how much energy is required to bring it from water at 0.00°C to vaporize it completely?

4. If the total amount of energy Christine expends during her run was given to the ice block, what would its final phase and temperature be?

Reflect

1. Was more heat used in changing the phase of the ice block to water or in changing its temperature from −1.00°C to 0.00°C?

2. Sketch a plot of T vs. Q for the ice block from −1.00°C to its final temperature.

240   Chapter 14

Section 14-7 Heat can be transferred by radiation, convection, or conduction

Goal: Describe the key properties of heat transfer by radiation, convection, and conduction.

Concept Check
1. Explain why hot air rises.

2. What are the quantities that determine the amount of heat an object loses by radiation? Explain.

3. What are the quantities that determine the amount of heat an object loses by conduction? Explain.

Problem-Solving Review
An electric stove has a copper coil that forms a circular disk of radius 8.00 cm and 0.500 kg. The pot of water is copper, has a base of the same area as the coil, and a thickness of 1.00 cm. The pot is filled with water at a temperature of 23.0°C. The coil is heated to 180°C.

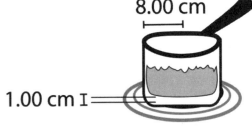

Set Up
1. Calculate the area of the coil in meters.

2. Write an expression for the rate of energy loss due to radiation by the coil.

Solve
1. What is the rate of energy loss due to radiation by the coil? (The emissivity of the coil is 0.060.)

2. Write an expression for the rate of heat flow from the coil to the water due to conduction in terms of the thermal conductivity of copper, the thickness and area of the pot, and the difference in temperatures between the pot and the coil.

3. Calculate the rate of heat flow from the coil to the pot due to conduction. (You can find the thermal conductivities of different materials in Table 14-5 of the textbook.)

Reflect
1. How would doubling the area of the coil affect the heat it produces by radiation?

2. Explain how a convection current can arise in boiling water.

Chapter 15 Thermodynamics II

Section 15-1 The laws of thermodynamics involve energy and entropy

Goal: Explain the general ideas of the laws of thermodynamics.

Concept Check

1. Explain the general purpose of the first law of thermodynamics.

2. When you bring a hot cup of coffee into a room, which law of thermodynamics explains why the coffee cools and the room warms up instead of the other way around?

Section 15-2 The first law of thermodynamics relates heat flow, work done, and internal energy change

Goal: Define and be able to apply the first law of thermodynamics.

Concept Check

1. You are boiling a pot of water on the stove and the steam starts to push the lid of the pot upward. If we take the steam in the pot as our system, is the work W being done by the steam on the lid positive or negative?

Problem-Solving Review

Kay is preheating an oven to make some delicious bread. The oven is electric and consumes 600 W. As the air heats up it expands and after 14 minutes pushes open the door doing 1000 J of work. Kay closes the door quickly before an appreciable amount of air escapes. Let's find the change in internal energy before an appreciable amount of air escapes.

Set Up

1. Is heat flowing into or out of the stove? What is the sign of Q?

2. Is the system doing work or being worked on? What is the sign of W?

Solve

1. Write down the first law of thermodynamics for this system.

2. How many joules of energy has the oven consumed in the last 14.0 minutes?

3. Calculate the change in internal energy after the hot air pushes the door open.

Reflect

1. What percentage of the energy that the oven consumes was used to open the door?

Thermodynamics II

Section 15-3 A graph of pressure versus volume helps to describe what happens in a thermodynamic process

Goal: Describe the nature of isobaric, isothermal, adiabatic, and isochoric processes.

Concept Check

1. What is an isobaric process?

2. What is an isothermal process?

3. What is an adiabatic process?

4. What is an isochoric process?

Problem-Solving Review

Consider the following graph which shows a process that one mole of a monatomic ideal gas undergoes:

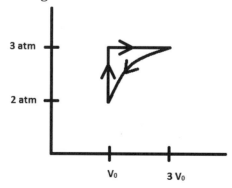

Set Up

1. What is the work done in the isobaric phase of the graph in terms of the pressure, V_f, and V_i. Calculate the work done assuming V_0 is 1.00 m³.

2. Write an expression for the work done in the isothermal phase of the graph ($T=250$K) in terms of N, T, V_f, V_i, and a constant. Calculate the work done.

3. Write an expression for work done by a system during an adiabatic process in terms of U.

4. Write the work done during the isochoric phase of the graph. Calculate the work done.

Solve

1. On a pV diagram sketch what an isobaric process looks like.

2. On a pV diagram sketch what an isothermal process looks like.

3. On a *pV* diagram sketch what a adiabatic process looks like.

4. On a *pV* diagram sketch what an isochoric process looks like.

Reflect

1. For each of your four sketches state whether the work done is positive, negative, or 0.

Section 15-4 More heat is required to change the temperature of an ideal gas isobarically than isochorically

Goal: Explain why different amounts of heat are required to change the temperature of a gas depending on whether the gas is held at constant temperature or constant pressure.

Concept Check
1. Explain the difference between specific heat c, molar specfic heat at constant pressure C_p, and molar specific heat at constant volume C_v.

2. Explain qualitatively why the value of C_v scales with the number of atoms in each gas molecule.

3. Explain why some quick thermodynamic processes are adiabatic.

Problem-Solving Review
Jim seals 1000 Moles of CO_2 in a test tube. He raises the temperature of the CO_2 from 23.0°C to 200.0°C. Let's calculate the thermodynamic properties of the system.

Set Up
1. Write an expression for the amount of heat flow needed to change the temperature under constant volume.

2. Write an expression for the molar specific heat at constant volume in terms of the number of degrees of freedom per molecule.

3. How many degrees of freedom does a molecule of CO_2 have at room temperature?

Solve

1. Calculate the molar specific heat at constant volume for CO_2. How does your answer compare to the experimental value given in Table 15-1 of the textbook.

2. How much heat is required to raise the temperature of the gas from 23.0°C to 200.0°C?

Reflect

1. How much more heat would it have taken if the gas were held at constant pressure instead of constant volume? (*Hint:* See Table 15-1 of the textbook to find the C_p of CO_2.)

Now consider the same test tube of CO_2, now with a movable plunger undergoing an adiabatic process. The CO_2 is again raised from 23.0°C to 200.0°C, but this time instead of being heated, the increase in temperature is accomplished by pushing down on the plunger very quickly. Let's calculate the change in pressure.

Set Up

1. What is the value of γ for CO_2? (*Hint:* You can either use the values of C_p and C_v you found above or you can look them up in Table 15-1 of the textbook.)

2. Find an expression relating pressure and volume for an ideal gas undergoing an adiabatic process.

3. Find an expression relating volume and temperature for an ideal gas undergoing an adiabatic process.

Solve

1. Use the expression you found in question 2 of the Set Up to find the ratio of initial pressure to final pressure in terms of γ and initial and final volumes.

2. Use the expression you found in question 3 of the Set Up to find the ratio of final pressure to initial pressure in terms of γ and initial and final volumes.

3. Combine your answers to questions 1 and 2 of the Solve section to find by what factor the pressure had to increase to cause the change in temperature.

Reflect

1. Would the change in pressure have been greater, smaller, or equal if the gas had been monotonic?

Section 15-5 The second law of thermodynamics describes why some processes are impossible

Goal: Define the second law of thermodynamics and its application to heat engines and refrigerators.

Concept Check
1. What is the second law of thermodynamics?

2. What are the three parts of a heat engine? Explain how a heat engine works in terms of these three parts.

Problem-Solving Review
One of the frustrating things about refrigerators is that the more you need them, the worse they work. Consider your refrigerator set at 3.0°C in the winter, where the indoor temperature in your house is 15.0°C, and the summer when it is 38.0°C.

Set Up
1. Sketch Q_H, Q_C, and W and their directions in a diagram for a refrigerator similar to Figure 15-14.

Solve
1. What is the maximum possible COP (coefficient of performance) for the refrigerator in the summer?

2. What is the maximum possible COP for the refrigerator in the winter?

3. If the refrigerator is rated at 350 W, what is the maximum energy it can remove in a second in the summer?

4. If the refrigerator is rated at 350 W, what is the maximum energy it can remove in a second in the winter?

Reflect

1. What would have to happen to T_H and T_C to get the COP as high as possible?

2. Why can't the refrigerator ever cool to absolute zero?

254   Chapter 15

Section 15-6 The entropy of a system is a measure of its disorder

Goal: Explain the concept of entropy and the circumstances under which entropy changes.

Concept Check

1. Explain under which conditions the entropy of a system could be made to decrease.

2. Explain the relationship between efficient engines and net entropy.

Problem-Solving Review

Marilyn places a piston on an electric stove with only 100 moles of air inside. The electric stove keeps the piston at a constant 150°C. Let's calculate the entropy of the air as it expands to twice its original volume.

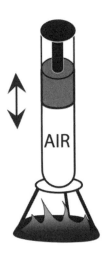

Set Up

1. What is the change in internal energy for an isothermal process?

2. Find an expression for the work done by an ideal gas in an isothermal process.

3. Write an expression for the entropy change in a reversible isothermal process.

Solve

1. Calculate the work done by the air as it expands to twice its volume.

2. How much heat enters the gas as it is expanding?

3. What is the change in entropy as the gas expands?

Reflect

1. What happened to the entropy in the reservoir that gave heat to this system?